HEIDELBERG SCIENCE LIBRARY | **Volume 17**

An Introduction to Macromolecules

Leo Mandelkern

The English Universities Press Ltd. London
Springer-Verlag New York • Heidelberg • Berlin

1972

Leo Mandelkern
Professor of Chemistry
The Florida State University
Tallahassee, Florida

© 1972 by Springer-Verlag New York Inc.

Library of Congress Catalog Card Number 72-83670.

Printed in the United States of America.

ISBN 0-340-17177-4 The English Universities Press Ltd. London.

ISBN 0-387-90045-4 Springer-Verlag New York • Heidelberg • Berlin

ISBN 3-540-90045-4 Springer-Verlag Berlin • Heidelberg • New York

To Berdie and Our Three Sons

Preface

The purpose of this volume is to provide beginning students in the natural sciences with a very elementary introduction to the structure and properties of macromolecular substances. Most students, either those who are science majors or those in a liberal art program, do not usually have the opportunity to learn about this important class of molecules until they take advanced courses at a senior or graduate level. There is thus a serious gap in our educational processes which has become acute with the widespread recognition of the important role played by macromolecules in biological systems, their extensive use in a myriad of articles of commerce, and their intrinsic scientific interest.

We have attempted to rectify this deficiency by presenting a detailed, but nonmathematical, description of all classes of macromolecules. Macromolecules are a unique class of substances in that they are molecules of extremely high molecular weight. When certain common structural features are recognized among the different kinds of macromolecules, they can be treated and analyzed from a unified point of view. With this approach, we can discuss in one volume macromolecules of biological interest as well as synthetic materials such as rubbers, glasses, coatings, and fibers.

It is our hope that this volume will appeal to students in introductory chemistry, biology, and physics courses and will open to them new vistas and horizons that they would not otherwise experience. It is important to emphasize that it is not intended to serve as an introduction to research in this field. If, however, we succeed in stimulating some readers to pursue further study in this area, we shall be extremely gratified and our efforts will have been well rewarded.

The generous assistance and advice from two colleagues is gratefully acknowledged. Professor G. R. Choppin read a portion of the manuscript in draft form and contributed some very expert advice as to the level and content. Professor K. B. Hoffman was kind enough to read the entire manuscript. Her suggestions and criticisms were invaluable. Without them it is doubtful whether the manuscript would have been completed. Finally, it is a pleasure to acknowledge Mrs. Suzanne Knuth and Mrs. Sandra Burkholder, who typed the manuscript and prepared it in final form. The author is indebted to them for their labors and for the skills that they brought to the task.

Tallahassee, Florida
February, 1972

Leo Mandelkern

Table of Contents

An Introduction to Macromolecules

Chapter I

INTRODUCTION

The history of mankind has been divided into a number of eras on the basis of the material common to the age. Thus, we have passed through the stone, copper, bronze, and iron ages. The modern epoch may very well be looked back on by historians as the Age of Macromolecules. Macromolecules, which are also known as polymers, provide both the necessities and amenities of life. Some macromolecules are found in nature in the form of wood, hide, wool, cotton, silk, and rubber. Others are synthetic, or man-made, and are used in our everyday life in the form of plastics, fibers, rubbers, packaging material, the paper upon which we write, paints, and coatings. Macromolecules such as carbohydrates, proteins, and nucleic acids are fundamental to the biological sustenance of life and are indeed the essence of life itself. Natural occurring polymers have of course been in use for a very long time. However, it is only recently that we have begun to understand the molecular structure of these materials and the relationship that exists between structure, properties, and function. With this new understanding we have been able to use the natural polymers more effectively and have also been able to devise synthetic materials to replace them. We can now even create new macromolecules with properties or combinations of properties that are not paralleled in nature. Thus, we are living in an age in which new materials are constantly replacing older ones. The potential and future benefit to mankind of this understanding appears to be unlimited.

A macromolecule (or "giant molecule" or polymer) is a chemical species of very high molecular weight. The word "polymer" comes from the Greek and means many parts. For example, water, H_2O, a very common low molecular weight substance, has a molecular weight of 18. This means that 1 mole of water, 6.06×10^{23} molecules, will weigh 18 grams. Benzene, a common organic solvent which is another low molecular weight substance, has a molecular weight of 78. However, 1 molecule of natural rubber has a molecular weight of about 1 million, i.e., 10^6. One mole of rubber will therefore weigh 10^6 grams, which is more than 1 ton. This comparison gives us a vivid picture of the difference between polymers and substances of low molecular weight. The high molecular weight of a polymer is a result of joining together in long chains, by means of chemical bonds, many thousands of atoms. The same types of covalent chemical bonds that are involved in the simpler low molecular weight organic molecules hold these giant molecules together. The kinds of atoms and their geometric arrangement give each macromolecule its chemical distinctiveness and thus its particular use and function.

The high molecular weight of these substances also reflects itself in their molecular size. For example, water has a molecular diameter of 4 Angstrom (Å) units, which corresponds to 4×10^{-8} cm. The diameter of the benzene molecule is only about 6 Å. The atoms in the protein hemoglobin, which has a molecular weight of 64,000, are so arranged that the diameter of this molecule is about 65 Å. A synthetic polymer such as polyethylene, having a molecular weight of 100,000, encompasses a very large volume. Its molecular diameter is about 640 Å. The molecular sizes of macromolecules are thus very much larger than what we have been accustomed to from our studies of low molecular weight species.

Although polymeric substances have been used by man for many centuries and although the structural principles of organic chemistry were developed in the 1850's, it is only recently that the macromolecular concept has been recognized and accepted. The possibility that covalent structures comprising thousands of atoms could exist as stable well-defined entities was not easily conceived. Furthermore, what we now recognize as polymers could not be readily analyzed by the prevailing classical methods of chemistry. In the second half of the nineteenth century and the early part of the twentieth, the study of polymers was virtually neglected since they could not be easily purified by the accepted methods of distillation and crystallization. They were thought of as a collection of small molecules held together by some peculiar and mysterious kinds of intermolecular forces not present in the simpler compounds of low molecular weight. In retrospect, it seems very strange indeed that it was not until the 1920's that serious investigations were undertaken of these substances, which are so essential to man's health, comfort, and well-being.

At about this time physical and physical-chemical methods of analysis were being developed and were achieving a respectability comparable to the more classical methods. Through the herculean efforts of a few pioneers, notably H. Staudinger in Germany, the macromolecular concept began to emerge and to be gradually, although in some instances grudgingly, accepted. By the mid-1940's the study of macromolecules was well on its way to becoming the quantitative science that it is today.

The acceptance of the concept that polymers are true molecular species of very high molecular weight is the primary reason that great advances have been made in the chemistry and physics of these substances in recent years. Because of the common structural feature of a large number of atoms covalently linked to one another, it is possible to examine the properties of all kinds of macromolecules from a unified point of view. At the same time one must be careful to recognize any distinguishing chemical features that exist in a given macromolecule.

In the following chapters we shall derive the benefits of these past endeavors as we examine how the basic principles of chemical bonding and structure manifest themselves as very specific properties and characteristics in macromolecular systems. As we shall see, there are a very large number of diverse properties, functions, and uses that can be realized. These range, for example, from the transmission of genetic information from one generation to another in living organisms to the development of articles of commerce as rubber-like materials which can function at very high temperatures. The uses of polymers are very widespread and varied. Before we embark on an analysis of the relations between structure and properties of macromolecular systems, it will be very helpful to classify the major kinds of polymers and to describe briefly the ways in which they are formed.

Chapter II

STRUCTURAL FEATURES AND PREPARATION

Classification and Definition

A polymer chain can be conveniently described by specifying the kind of repeating units present and their spatial arrangement. Molecules composed of chemically and stereochemically identical units (terminal units excepted) are termed homopolymers. When the chain is composed of more than one kind of repeating unit, it is called a copolymer. The chain units can be interconnected in a variety of ways that must, of course, be consistent with the rules of chemical bonding. It is therefore possible for many different geometrical patterns to evolve. The simplest polymer, in terms of chain structure, is the linear homopolymer. As the name implies, the chain units, in this case, are arranged in linear sequence that can be represented schematically as shown by Fig. 1. Here

Y———A—A—A—A—A—A—A—A—A—A—A—A—A———X

Fig. 1 Linear homopolymer.

A represents the repeating or structural unit, and *X* and *Y* are the terminal units that may or may not be different from one another and from *A*.

Departures from this simple linear array lead to structures of increasing geometric complexity. A nonlinear or branched structure such as shown in Fig. 2 can easily be formed. Here, neither the length of the

Fig. 2 Branched polymer.

branches nor their distribution along the main chain has to be uniform. However, for branching to occur, some of the units must possess bonding characteristics that are different from the predominant repeating unit. These differences, however, need not be very large, as is seen in the following comparison of the structure of linear and branched polyethylene (Figs. 3 and 4). The chemical difference between these two molecu-

Fig. 3   Linear polyethylene.

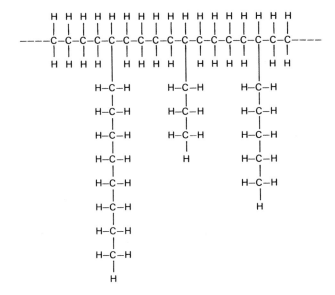

Fig. 4   Branched polyethylene.

lar chains is obviously quite small. Yet, because of the structural differences, these two molecules have quite different properties. The relationships between the structure and properties will be discussed in detail in subsequent chapters. Despite their chemical identity, the uses that can be made of these two polymers is, therefore, quite different.

Branched structures are not necessarily restricted to the particular form of Fig. 2. A more extensive distribution of branched points results in very highly ramified shapes. Thus, structures such as that shown in

5

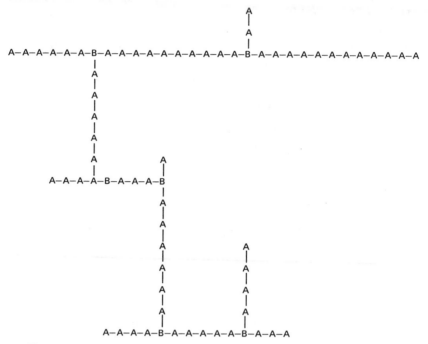

Fig. 5 Highly branched polymer.

Fig. 5 can develop. A continuation of this pattern will lead to a very complicated three-dimensional structure, illustrated in Fig. 6. Although for branched chains the concept of discrete molecules can still be maintained, it has to be abandoned when considering a three-dimensional or infinite network. A given network encompasses the complete macroscopic domain of the sample. Hence, the expression infinite network.

Nonlinear and network structures can also be prepared from a collection of linear chains by covalently linking together chain units selected from different molecules. Such a system is said to be cross-linked and can be represented as shown in Fig. 7. When a sufficient number of units are intermolecularly cross-linked, then an infinite network will be formed.

So far we have concentrated on the geometry of structures that can be formed. In some cases a very small fraction of chemically different units has been introduced into the chain. Linear chains can also be made up of different kinds of repeating units. Such chains are called linear copolymers. The properties of a copolymer depend not only on the chemical nature and amounts of the counits, but also very strongly

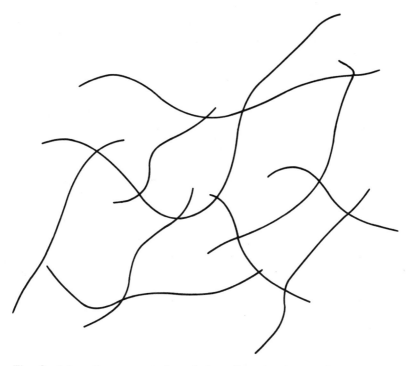

Fig. 6 Schematic representation of three-dimensional network.

Fig. 7 Cross-linked system.

on how the units are distributed along the chain. For example, given two distinctly different kinds of repeating units forming a chain of fixed composition, there are a large number of different ways in which these units can be arranged in a linear sequence. Three extreme possible arrangements can be classified conveniently. These are an alternating copolymer of A and B such as shown in Fig. 8. In this case there is a very strong propensity for a unit to be followed by one of the opposite kind.

‒‒‒A‒B‒A‒B‒A‒B‒A‒B‒A‒B‒A‒B‒A‒B‒A‒B‒A‒B‒A‒‒‒

Fig. 8 Alternating copolymer.

‒A‒A‒A‒A‒A‒A‒A‒B‒B‒B‒B‒B‒B‒B‒A‒A‒A‒A‒B‒B‒B‒B‒B‒A‒A‒A‒A‒A‒‒‒

Fig. 9 Ordered copolymer.

At the opposite extreme is the block or ordered copolymer (Fig. 9), where there is an overwhelming tendency for a unit to be succeeded by another of the same kind. Here long sequences of one type of unit alternate with long sequences of the other kind. The third major classification is a random copolymer, where the different units are randomly distributed along the chain. Such a linear structure can be represented as shown in Fig. 10. It is a tribute to the skills of the synthetic polymer

‒‒B‒A‒A‒B‒A‒A‒B‒B‒A‒A‒A‒B‒A‒A‒B‒B‒A‒A‒A‒B‒A‒B‒B‒A‒A‒B‒‒

Fig. 10 Random copolymer.

chemist that from the same set of counits these different kinds of chains, with correspondingly different properties, can be produced.

Chemical Repeating Units

An exhaustive listing of macromolecules that have been prepared in the laboratory and found to be useful, as well as those that occur naturally and are involved in biochemical and biological function, would run to many hundreds of entries. Such a compilation would be well beyond our present capacity and interest. Instead, we list in Table I some typical linear homopolymers which were selected to illustrate the tremendous chemical versatility that is found in macromolecular systems.

Many of the homopolymers we encounter daily are listed in Table I along with the chemical formula of the repeating unit, their proper chemical name, and in many cases their common name. Strictly speaking, some of the homopolymers listed could be thought of as perfectly alternating copolymers. However, since there is an exact repeat of the structural unit, it is preferable to classify them as homopolymers. Included in this list are naturally occurring polymers such as gutta-percha, natural rubber, and cellulose and its derivatives, which together with the syn-

Table I

Homopolymers

Repeating Unit	Poly —	Common or Generic Name
$\left(\begin{array}{c} H\ \ H \\ \vert\ \ \ \vert \\ -C-C- \\ \vert\ \ \ \vert \\ H\ \ H \end{array}\right)$	Ethylene	
$\left(\begin{array}{c} H\ \ H \\ \vert\ \ \ \vert \\ -C-C- \\ \vert\ \ \ \vert \\ H\ \ CH_3 \end{array}\right)$	Propylene	poly olefin
$\left(\begin{array}{c} H\ \ CH_3 \\ \vert\ \ \ \vert \\ -C-C- \\ \vert\ \ \ \vert \\ H\ \ C=O \\ \vert \\ OCH_3 \end{array}\right)$	Methyl methacrylate	plexiglas, lucite
$\left(\begin{array}{c} H\ \ H \\ \vert\ \ \ \vert \\ -C-C- \\ \vert\ \ \ \vert \\ H\ \ Cl \end{array}\right)$	Vinyl chloride	vinyl
$\left(\begin{array}{c} H\ \ H \\ \vert\ \ \ \vert \\ -C-C- \\ \vert\ \ \ \vert \\ H\ \bigcirc \end{array}\right)$	Styrene	
$\left(\begin{array}{c} H\ \ H \\ \vert\ \ \ \vert \\ -C-C- \\ \vert\ \ \ \vert \\ H\ \ CN \end{array}\right)$	Acrylonitrile	acrylic, orlon
$\left(\begin{array}{c} F\ \ F \\ \vert\ \ \ \vert \\ -C-C- \\ \vert\ \ \ \vert \\ F\ \ F \end{array}\right)$	Tetrafluoroethylene	teflon
$\left(\begin{array}{c} H\ \ CH_3 \\ \vert\ \ \ \vert \\ -C-C- \\ \vert\ \ \ \vert \\ H\ \ CH_3 \end{array}\right)$	iso-Butylene	butyl rubber

Table I (continued)

Repeating Unit	Poly —	Common or Generic Name
$\begin{pmatrix} CH_3 & H \\ \backslash & / \\ C = C \\ / & \backslash \\ CH_2 & CH_2 \end{pmatrix}$	*cis*, 1,4 isoprene	natural rubber
$\begin{pmatrix} & H_2 \\ CH_3 & C \\ \backslash & / \\ C = C \\ / & \backslash \\ CH_2 & H \end{pmatrix}$	*trans.* 1,4 isoprene	gutta percha
$\begin{pmatrix} CH_2 & H \\ \backslash & / \\ C = C \\ / & \backslash \\ Cl & CH_2 \end{pmatrix}$	*trans,* 1,4 chloroprene	neoprene
$\begin{pmatrix} CH_2 & H \\ \backslash & / \\ C = C \\ / & \backslash \\ H & CH_2 \end{pmatrix}$	*trans,* 1,4 butadiene	
$\begin{pmatrix} H \\ \| \\ C-O \\ \| \\ H \end{pmatrix}$	Methylene oxide	polyether, delrin
$\begin{pmatrix} CH_3 \\ \| \\ Si-O \\ \| \\ CH_3 \end{pmatrix}$	Dimethyl siloxane	silicone
$\begin{pmatrix} O & & O & H & & H \\ \| & & \| & \| & & \| \\ C-(CH_2)_4-C-N-(CH_2)_6-N \end{pmatrix}$	Hexamethylene Adipamide	nylon (6,6)
$\begin{pmatrix} H & H & & & H & H & H & H \\ \| & \| & & & \| & \| & \| & \| \\ C-C-O-C-C-C-C-C-C-O \\ \| & \| & & \| & \| & \| & \| & \| & \| \\ H & H & & O & H & H & H & H & O \end{pmatrix}$	Ethylene Adipate	polyester

Table I (continued)

Repeating Unit	Poly —	Common or Generic Name
	Ethylene terephthalate	dacron, mylar
	Ethylene isocyanate	poly urethane
	Bisphenol-A-carbonate	poly carbonate, lexan
	Cellulose	cotton, wood, viscose rayon, cellophane
	Cellulose acetate[1]	acetate rayon
	Cellulose nitrate[1]	celluloid, smokeless powder

Table I (continued)

Repeating Unit	Poly —	Common or Generic Name
	Amylose[2]	starch
	Poly-alanine	polypeptide, poly α-amino acid
	Poly adenylic acid	nucleic acid

[1] The partially substituted cellulose derivatives are listed. When completely substituted the compounds are called cellulose triacetate and trinitrate respectively.

[2] Chemical studies have shown that there are two kinds of chain structures in starch each having the same composition. The linear chain is called amylose; the branched chain amylopectin.

thetic polymers are widely used in a variety of articles of commerce. We also have listed chain molecules such as starch, which is a polysaccharide, polypeptides, and nucleic acids. These can be organized into more complex macromolecular structures of great biological and biochemical importance. We shall reserve a more detailed discussion of macromolecular systems of biological importance for Chapter VI. Because of the different kinds of repeating units that can be used and the varied nature of the functional groups that can be attached to the main chain atoms, a great deal of chemical diversity can be built into a polymer chain.

The list of homopolymers in Table I gives a vivid demonstration of how the chemical nature of the repeating unit influences properties since the polymers have many different end-uses. Before pursuing this point, however, it is instructive to examine some of the differences between the properties of a long-chain molecule and a similarly constituted monomeric substance. The properties of the linear long-chain hydrocarbon, polyethylene, which can be represented by a string of CH_2 units, as in Fig. 3, serves as a good example. The primitive monomeric analogue

is methane, $H-\underset{\underset{\displaystyle H}{|}}{\overset{\overset{\displaystyle H}{|}}{C}}-H$, which is a gas of molecular weight 16. The chain can

then be built up by forming successively, ethane $H-\underset{\underset{\displaystyle H}{|}}{\overset{\overset{\displaystyle H}{|}}{C}}-\underset{\underset{\displaystyle H}{|}}{\overset{\overset{\displaystyle H}{|}}{C}}-H$, propane

$H-\underset{\underset{\displaystyle H}{|}}{\overset{\overset{\displaystyle H}{|}}{C}}-\underset{\underset{\displaystyle H}{|}}{\overset{\overset{\displaystyle H}{|}}{C}}-\underset{\underset{\displaystyle H}{|}}{\overset{\overset{\displaystyle H}{|}}{C}}-H$, and butane $H-\underset{\underset{\displaystyle H}{|}}{\overset{\overset{\displaystyle H}{|}}{C}}-\underset{\underset{\displaystyle H}{|}}{\overset{\overset{\displaystyle H}{|}}{C}}-\underset{\underset{\displaystyle H}{|}}{\overset{\overset{\displaystyle H}{|}}{C}}-\underset{\underset{\displaystyle H}{|}}{\overset{\overset{\displaystyle H}{|}}{C}}-H$, which also are gases. As the

number of carbon atoms in the linear array is increased to six, a very volatile liquid, n-hexane, is formed. As the carbon number is increased, the viscosity of the liquid also increases. When 18 carbon atoms are introduced into the chain, a solid wax-like substance, or normal paraffin, results. The molecular weight of the compound formed at this point is 254. In this way, one can build up a long chain of very high molecular weight. Polyethylenes, comprised of hundreds of thousands of chain carbon atoms and thus having molecular weights in the millions, have been synthesized. Surprising as it may seem, some of the major structural features of the low molecular weight normal paraffins can still be found in these very long chains.

When some of the hydrogen atoms from polyethylene are systematically replaced by other atoms or groups of atoms, the chemical changes that are made also affect the physical properties. The relationships between chemical constitution, structure, and physical properties of long chain molecules are discussed in detail in Chapters IV and V. As an example, if all the hydrogen atoms are replaced by fluorine atoms, a very interesting polymer, polytetraflouroethylene, popularly called Teflon, results. The properties of this polymer are quite different from those of its hydrogen counterpart. It is very inert chemically and resists attack by virtually all chemical reagents. In addition, it is one of the most thermally stable polymers known and does not undergo any physical or chemical changes below temperatures of about 320°C. This polymer possesses a very low coefficient of friction, which makes it very useful as a coating. Teflon also has the property that other materials do not adhere to it so that it presents a "nonsticky" surface.

Table I lists two sets of naturally occurring polymers. These are the *cis* and *trans* poly 1,4 isoprenes and the polysaccharides starch[1] and cellulose. In each of these groupings the two species have identical chemical formulas. *Cis*-polyisoprene, commonly called natural rubber, displays the expected elastic properties associated with the name. On the other hand, the *trans* polymer, gutta-percha, is a hard and inelastic substance at room temperature. The only difference between the two chains is the geometrical arrangement about the double bond as is shown in Fig. 11. For the *cis* polymer the CH_2 groups lie on the same side of the

Trans polyisoprene — gutta percha

Cis polyisoprene — natural rubber

Fig. 11 Skeletal structure of *cis* and *trans* polyisoprene.

double bond as the chain is perpetuated. For the *trans* polymer the CH_2 groups are arranged on alternate sides of the double bond. Although this difference appears to be only a minor structural alteration, it manifests itself in a set of very different properties.

---

[1] Although we are using the word "starch," we are in fact referring to the linear amylose chain.

14

Similarly, although starch and cellulose chains are made up from the same building block, the glucose unit, they have quite different properties. Starch is readily soluble in many solvents, including water. It is an important foodstuff that is easily digestible by humans, and in the body the polymer is degraded to the monomeric sugar by enzymatic action. On the other hand, cellulose, a major constituent of plants, is a very insoluble substance and is not digestible by man. However, animals such as cows and horses possess the enzymatic apparatus necessary to degrade cellulose to its monomeric unit. The termite plays havoc with our houses by being able to digest and make a meal of wood (cellulose). The reason for these remarkable differences resides again in the geometric nature of the linkage propagating the chain. Fig. 12 gives a perspective

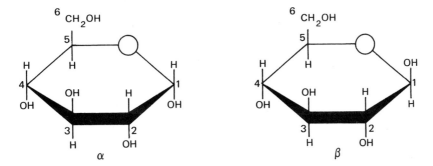

Fig. 12 Two configurations of the glucose unit.

drawing of the glucose unit. The ring is perpendicular to the plane of the paper with the heavily marked bonds pointing outwards. The oxygen atom is then directed to the rear with the ring substituents directed above or below the plane. The two OH groups attached to the carbon atoms in the 1 and 4 positions can either be on the same side of the plane or be opposite to one another. The linkage that perpetuates the chain, the ether type oxygen bond, can be thought of as resulting from the reaction of the 1 and 4 hydroxyl groups from different rings. The relative position of these two groups in the monomers thus determines an important distinguishing feature of the chain as is shown in Fig. 13. The α-linkage characteristic of starch is always directed below the plane while the β-linkage of cellulose is directed above the plane. This structural difference is, in turn, reflected in the diverse chemical behavior and physical properties of the two chains. Each of these naturally occurring polymers is always found with its characteristic linkage throughout the chain. A chain with mixed linkages has not as yet been found.

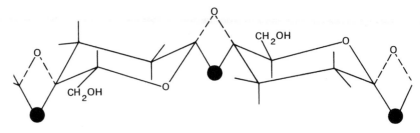

Fig. 13 Chain structure of starch and cellulose. The α-linkage of starch is shown by the solid lines. The β linkage of glucose by the dashed lines.

Copolymers

A further illustration of the chemical virtuosity that can be attained by long chain molecules manifests itself in copolymers. In Table II we list some of the synthetic copolymers that have become important.

Table II

Copolymers

Butadiene/Styrene 3:1 Random	Synthetic Rubber; GRS, SBR
Ethylene/Propylene 3:1 Random	Elastomer
Acrylonitrile/Butadiene/Styrene Random	ABS; tough plastic, high impact strength
Ethylene/Vinyl Acetate Random	Elvax; wax coatings
Vinyl chloride/vinyl acetate 8:1 Random	Vinylite; coatings
Vinyl chloride/acrylonitrile 9:1 Random	Dynel; fibers, wigs
Acrylonitrile/vinyl acetate 8.5:1 Random	Acrylic fiber; Orlon, Acrilan
Vinylidene chloride/vinyl chloride Random	Saran; packaging, fibers
Vinylidene fluoride:hexafluoropropylene 7:3 Random	High temperature rubber
Polyurethane/polyether Block	Spandex; elastic fiber
Styrene/butadiene Block	Vulcanized rubber at room tempera- ture; undergoes plastic flow at high temperatures; can be molded

As we have mentioned, the properties of copolymers depend not only on chemical constitution but also on the sequential distribution of counits along the chain. For example, with an ordered copolymer desirable properties characteristic of each of the corresponding homopolymers can be incorporated into a single molecular chain. The same properties cannot be obtained, however, merely by mechanically mixing the individual homopolymers. Thus, for example, in the "spandex type" of copolymer the apparently contradictory properties of simultaneously having fiber-like qualities, with the associated high tensile strength, while enjoying the benefits of long-range elasticity are incorporated into one molecular system. Similarly, block copolymers can be prepared that enhance the dyeing capacity of a fiber while still maintaining the other necessary physical properties.

The properties of random and alternating copolymers, on the other hand, are usually quite different from those of the corresponding homopolymers. For example, the random ethylene-propylene copolymer is an elastomer. It has very good resistance to oxidation because of the absence of double bonds, which are found in most other rubber-like polymers. The elastomeric quality of this copolymer is not found in the corresponding homopolymers, polyethylene and polypropylene. These two homopolymers are partly stiff, plastic materials. The butadiene-styrene copolymers, in a 3:1 mole ratio, are also elastomers and are particularly suited for heavy duty automotive use. As the relative amount of styrene is increased, however, the material becomes tough and leathery. In contrast, a block copolymer of styrene and butadiene, although behaving like a vulcanized rubber at room temperature, undergoes plastic flow at high temperatures and can be easily molded into any desired form. The random copolymerization of three species—acrylonitrile, butadiene, and styrene—yields a very tough, rigid material, ABS. Although many polymeric materials are rigid and hard, they are usually also very brittle. Because of the special combination of monomers, this copolymer possesses unique mechanical properties in that it is not only rigid and tough but also has a high impact strength. Hence, it does not fracture very easily. ABS, therefore, has many of the properties associated with a metal, but it is also very lightweight and corrosion resistant. It finds a variety of uses, from football helmets to living room furniture.

Acrylic type fibers, which are based on polyacrylonitrile, have other units such as vinyl acetate randomly introduced into the chain. The necessity for a copolymer here is to facilitate the polymer processing and to make the fiber more amenable to dyeing with conventional textile dyes. Vinyl acetate is also a common copolymeric unit introduced into polyvinylchloride. Many fabrication problems are alleviated by this procedure, and the resulting physical properties are improved.

The random copolymerization of vinylidene fluoride with hexafluoropropylene yields a substance whose mechanical properties are quite different from either of the parent homopolymers. Instead of the rigid, nondeformable material associated with fluorine-containing homopolymers, this copolymer possesses very good elastomeric qualities. At the same time, however, the thermal stability with respect to chain degradation, which is a special characteristic of fluorine-containing polymers, is maintained. That is to say, the integrity of the chain is maintained so that this copolymer behaves as a rubber at very high temperatures. Most of the common elastomers cannot be used at elevated temperatures because of chemical degradation.

The few examples cited should demonstrate how control of physical and chemical properties can be achieved by appropriate copolymerization when carried out by skilled synthetic chemists.

Preparation

We have been discussing the architectural forms that long-chain molecules can adopt, the great diversity in the chemical types of repeating units that are now available, and in a very general and qualitative manner, the relationship of these factors to properties. We have not as yet considered, however, the processes that enable these giant molecules to be synthesized, either in the laboratory or in nature, from their monomeric components. In discussing polymerization we shall be concerned primarily with the main principles that are involved, and we shall not dwell on any detailed mechanism of a particular reaction.

It is readily apparent and indeed axiomatic that in order to perpetuate a chain to appreciable size, the functionality of the starting monomeric species must be at least two. The meaning of this statement can be seen in the following example. Let us consider two species (a) R—A and (b) R'—B that react with one another to form a third species (c) R—A—B—R' Thus

$$R-A + B-R' \longrightarrow R-A-B-R' \tag{1}$$

and the new compound (c) is formed by the combination of groups A and B. No further reaction of the same type involving species (c) can take place since there are no longer any reactive groups available. Thus, one molecule of species (a) can react once and only once with a molecule of species (b) to give (c). Compounds (a) and (b) are thus said to be monofunctional. A simple example of this scheme, from organic chemistry, is seen in Eq. (2) for the reaction of an acid (1) with an alcohol (2) to give an ester (3) and water,

$$CH_3-\overset{\overset{\displaystyle O}{\|}}{C}-OH + HO-CH_2-CH_3 \longrightarrow CH_3-\overset{\overset{\displaystyle O}{\|}}{C}-O-CH_2CH_3 + H_2O \tag{2}$$

(1)   (2)   (3)

Here, as is indicated schematically, the carboxyl group of the acid,
$\overset{O}{\overset{\|}{C}}$—OH, has reacted with the hydroxyl group, OH of the alcohol to form
the ester. In the process, water is given off. In this case both the acid
and the alcohol are monofunctional. If either the acid or the alcohol were
bifunctional so that, for example, the acid contained two carboxyl groups
while the alcohol contained only one hydroxyl group, the reaction scheme
indicated by Eq. (2) could proceed one more step. The product of the
reaction would, however, still be a low molecular weight species.

Let us now see what happens when each of the reactants is bifunctional.
Therefore, (a) =A—R—A and (b) =B—R'—B. Now, each of these species has
two groups that are capable of reacting with one another. For our schematic representation of the reaction we can write

$$A-R-A + B-R'-B \longrightarrow A-R-A-B-R'-B \tag{3}$$
$$\text{(a)} \qquad \text{(b)} \qquad\qquad\qquad \text{(c)}$$

We now find, contrary to our previous experience, that the product
is also bifunctional. Hence further reaction, with the concomitant building up of the molecule or chain, can take place. For example, we can
have

$$A-R-A + A-R-A-B-R'-B \longrightarrow A-R-A-B-R'-B-A-R-A \tag{4}$$
$$\text{(a)} \qquad\quad \text{(c)}$$

The new species formed is still bifunctional and could react further
with (b). Another possibility would be for (c) to react with itself so
that

$$A-R-A-B-R'-B + A-R-A-B-R'-B \longrightarrow A-R-A-B-R'-B-A-R-A-B-R'-B \tag{5}$$
$$\text{(c)} \qquad\qquad\qquad \text{(c)}$$

Here again the bifunctionality is maintained in the product so that further reaction can take place. Therefore, with each successive step more
reaction pathways develop. These become much too numerous to attempt
to describe in detail. However, it should be abundantly clear that in
this stepwise manner very long chain molecules can be formed. The
length of each chain will not be the same and consequently there will
be a distribution of molecular sizes in the resulting product.

Many important classes of polymers are actually synthesized by the
method just outlined. This mode of synthesis is called condensation polymerization. For example, in analogy with Eq. (3), if our initial reactants
are a dialcohol such as ethylene glycol, $HOCH_2 CH_2 OH$, and a diacid such

as adipic acid, HO–$\overset{\displaystyle O}{\overset{\|}{C}}$–$\overset{\displaystyle H}{\underset{\displaystyle H}{\overset{\|}{\underset{\|}{C}}}}$–$\overset{\displaystyle H}{\underset{\displaystyle H}{\overset{\|}{\underset{\|}{C}}}}$–$\overset{\displaystyle H}{\underset{\displaystyle H}{\overset{\|}{\underset{\|}{C}}}}$–$\overset{\displaystyle H}{\overset{\|}{C}}$–$\overset{\displaystyle O}{\overset{\|}{C}}$–OH, in the first step of the

polymerization,

$$HO-\overset{O}{\overset{\|}{C}}-CH-CH-CH-CH-\overset{O}{\overset{\|}{C}}-OH \ + \ HO\,CH-CH-OH \ \longrightarrow$$

$$HO-CH-CH-O-\overset{O}{\overset{\|}{C}}-CH-CH-CH-CH-\overset{O}{\overset{\|}{C}}-OH \qquad (6)$$

$$+ \ H_2O$$

water is given off and an ester is formed. This ester contains an OH (hydroxyl) and a $\overset{O}{\overset{\|}{C}}$–OH (carboxyl) end group. Further stepwise reactions can take place, and a polymer will be formed. In this example a linear polyester, of the type given in Table I, is synthesized. When a diamine such as hexamethylene diamine H–$\overset{\displaystyle H}{\underset{\displaystyle H}{\overset{\|}{\underset{\|}{N}}}}$–$\overset{\displaystyle H}{\underset{\displaystyle H}{\overset{\|}{\underset{\|}{C}}}}$–$\overset{\displaystyle H}{\underset{\displaystyle H}{\overset{\|}{\underset{\|}{C}}}}$–$\overset{\displaystyle H}{\underset{\displaystyle H}{\overset{\|}{\underset{\|}{C}}}}$–$\overset{\displaystyle H}{\underset{\displaystyle H}{\overset{\|}{\underset{\|}{C}}}}$–$\overset{\displaystyle H}{\underset{\displaystyle H}{\overset{\|}{\underset{\|}{C}}}}$–$\overset{\displaystyle H}{\overset{\|}{N}}$–H is substituted for the dialcohol, a polyamide will be formed following the same synthetic scheme. Here water is formed by the reaction of H from the amine with the OH group of the acid. When a dialcohol and a diisocynate, $O{=}C{=}N{-}R{-}N{=}C{=}O$, react, then a polyurethane is synthesized. A very important concept is embodied in the above discussion. We have found, in effect, that the principles of organic chemistry that are involved in synthesizing the polymers are exactly the same as those applicable to monomeric systems.

Another major method of synthesis is termed addition polymerization. It is the mode by which ethylene $H_2C{=}CH_2$, propylene $H_2C{=}CHCH_3$, and vinyl monomers such as styrene $H_2C{=}CH\langle\!\bigcirc\!\rangle$ are polymerized to their respective homopolymers. When we look at the structure of the ethylene monomers, which can serve as a prototype for the others, $\overset{H}{\underset{H}{\diagdown}}C{=}C\overset{H}{\underset{H}{\diagup}}$ it would appear at first glance, that the necessary requirement of bifunctionality is not met. However, a more detailed examination and insight into the nature of the carbon-carbon double bond shows that this discrepancy is more apparent than real. The two electron pairs that form the double bond are not of the same strength.[2] One pair of electrons, or one bond, will be much more reactive than the other. Hence, although it is not obvious in its chemical formula, our monomer inherently possesses bifunctional character. To take advantage of the carbon-carbon double bond of monomers a catalyst is required for an initiation step. A very convenient catalyst is a free radical. This is a species with an unshared electron which can be represented as R·. Schematically, the polymerization proceeds as follows: In the first step, initiation, a monomer unit has

$$
\text{R} \cdot \ + \ \begin{matrix} \text{H} & & \text{H} \\ \backslash & & / \\ & \text{C} :: \text{C} & \\ / & & \backslash \\ \text{H} & & \text{H} \end{matrix} \ \longrightarrow \ \text{R} : \overset{\displaystyle \text{H}}{\underset{\displaystyle \text{H}}{\text{C}}} : \overset{\displaystyle \text{H}}{\underset{\displaystyle \text{X}}{\text{C}}} \cdot \tag{7}
$$

become a free radical. It, in turn, can now react with other monomer units to form a polymer by the following scheme:

$$
\text{R} - \overset{\displaystyle \text{H}}{\underset{\displaystyle \text{H}}{\text{C}}} - \overset{\displaystyle \text{H}}{\underset{\displaystyle \text{X}}{\text{C}}} \cdot \ + \ \begin{matrix} \text{H} & & \text{H} \\ \backslash & & / \\ & \text{C} :: \text{C} & \\ / & & \backslash \\ \text{H} & & \text{X} \end{matrix} \ \longrightarrow \ \text{R} - \text{C} - \text{C} - \text{C} - \text{C} \cdot \tag{8}
$$

The initiating or activating species is said to have opened up the double bond. As this propogation step, Eq. (8), is continued, hundreds to thousands of monomer units can be added to the growing chain. The center of activation or reactive site is at the end of the growing chain. Termination of the growing chain will occur when two free radicals combine. There are also other possible modes of termination. Thus, addition polymerization is essentially a chain reaction. In the case illustrated it proceeds by a free radical mechanism. By using different kinds of initiators the polymerization can also proceed by an ionic mechanism.

There are some very important differences between the two major polymerization processes we have been discussing. In condensation polymerization the reaction between two growing polymer chains leads to a further enhancement of the molecular weight. But in addition polymerization reaction between two growing chains leads to a cessation of further growth as far as the two molecules involved are concerned. In addition polymerization an individual polymer molecule, once initiated, is synthesized very rapidly from the unreacted monomer pool. However, to obtain a good total yield of polymer requires several hours. Very high molecular weights can be obtained quite routinely from this type of chain reaction. In condensation polymerization, the molecular weight

[2] A student of orbital theory will recognize that we are discussing σ and π bonds.

is built up by a series of independent reactions proportioned over the total time period of the polymerization. At the early stages of the process the average molecular weight of the system will be low, although the amount of unreactive material present will be negligible. As the reaction continues, the average molecular weight will also increase. However, to obtain very high molecular weights, the reaction will have to proceed virtually to completion. Molecular weights greater than 25,000 are rare for linear condensation polymer. In contrast, with addition polymerization very high molecular weights are found at the very early stages of the process. Continuation of addition polymerization is dictated by the amount of polymer required rather than an effort to obtain higher molecular weights.

Branching, Cross-Linking, and Network Formation

The polymerization schemes that have been outlined lead to the formation of linear chains in the absence of any complicating side reactions. To form branched structures, at least one of the monomeric units must have a functionality greater than two. If, for example, a trifunctional acid reacted with a difunctional alcohol, a highly ramified structure, of the kind illustrated in Fig. 5, would result. Many nonlinear polymers of interest are composed mainly of bifunctional units. The units with a higher functionality are usually a minor constituent and randomly distributed. In addition polymerization, branching and network formation can also be accomplished by including a small amount of

a divinyl monomer, such as divinyl benzene $H_2C\!=\!CH\!-\!\!\left\langle\bigcirc\right\rangle\!-\!CH\!=\!CH_2$

in the polymerization mixture.

Celluloid, which is nitrated cellulose, is generally credited with being the first plastic. A naturally occurring macromolecule is the starting material in this case. However, since celluloid is highly inflammable, a great deal of effort was made to find a substitute material. The first completely synthetic plastic was prepared by Baekeland in 1907 and is called Bakelite. It still has a widespread use despite the hundreds of polymers that have been synthesized since. Bakelite is not, however,

a linear chain. It is prepared from the reaction of phenol $\left\langle\bigcirc\right\rangle\!-\!OH$ with

formaldehyde $H\!-\!\overset{\overset{\displaystyle H}{|}}{C}\!=\!O$ to yield a three-dimensional network polymer as is indicated in Fig. 14. This polymer is a very strong and tough, but lightweight, material that is easily molded into a variety of objects.

22

Fig. 14 Bakelite-phenol-formaldehyde three-dimensional network.

A similar kind of highly cross-linked three dimensional network is formed by the reaction of formaldehyde and urea, $O=C\begin{smallmatrix}NH_2\\NH_2\end{smallmatrix}$, yielding a polymer with the structure as indicated in Fig. 15. The reaction of

Fig. 15 Urea-formaldehyde polymer.

formaldehyde with the trifunctional amine melamine

also yields a highly cross-linked network polymer. The melamine polymers are used extensively in lightweight dinnerware and as laminates for counter and table tops.

Three-dimensional networks can be formed from a collection of linear chains by the introduction of intermolecular cross-links after the polymers have been formed. This procedure requires having reactive groups in the chain and a cross-linking reagent that has at least bifunctional character, so that it can react with at least two such units. More than a century ago, Charles Goodyear discovered the vulcanization of rubber, which is the classical intermolecular cross-linking process. In Chapter IV we discuss in detail the importance of this chemical process to the elastic properties of natural rubber. The tanning of leather is merely the intermolecular cross-linking of its major constituent which is the fibrous protein collagen. Here the cross-linking process increases the thermal stability, reduces the shrinkage, and renders the material less susceptible to the action of solvents, particularly water. Wool and hair owe many of their physical properties to the presence of sulfur-sulfur intermolecular cross-links between polypeptide chains.

Another method of cross-linking, which is applicable to many polymeric systems, involves subjecting the chains to the action of high energy ionizing radiation from neutrons or gamma rays. In the case of polyethylene, for example, this high energy source strips an occasional hydrogen atom from the chain and leaves a free radical behind. The free radicals from different chains can combine with each other to form intermolecular covalent cross-links. This method of cross-linking has been used extensively with polyethylene. Since the cross-linking process retards flow, objects can be produced that retain their shapes at much higher temperatures than would be usual. This is an extremely useful property for branched polyethylene when sterilization in boiling water is necessary.

In this introductory and brief survey we have seen the diversity of structures and chemical types that can be manifested in macromolecules. Concomitantly, there is a wide range in properties that can be attained. These, in turn, direct and govern the end-use of synthetic polymers and the biological function of the naturally occurring ones. Although we have accepted the fact that such a wide range in properties exists, we have not as yet inquired as to what factors cause the differences. Put another way, the question can be asked as to why one type of chain makes a good elastomer while another is a tough inextensible plastic. Why can genetic information be stored in nucleic acids? Or why does nylon make such a good fiber comparable to nature's own silk and wool? We seek the answers to these and similar questions in the following chapters by first examining the structure of polymer chains in more detail and then relating the structure to properties.

Chapter III

CHAIN STRUCTURE

Internal Rotation

The pioneering work, over the past 30 to 40 years by such scientific luminaries as Staudinger, Svedberg, Debye, and Flory has established beyond doubt that molecules of very high molecular weight exist. By use of modern experimental and theoretical methods, a quantitative description that is the basis for the understanding of the properties of such molecules has been developed. The usual or normal laws of chemistry and physics were found to apply equally well to this class of molecules. However, in many cases, a generalization of the existing laws was necessary in order to include monomeric and polymeric systems in a unified way. This unification has led in fact to a better understanding of the underlying basis for these laws.

Establishing the fact that we are dealing with molecules of very high molecular weight is obviously the first step to an understanding of their properties. However, this knowledge is not in itself sufficient to explain the diversity of properties characteristic of macromolecular systems. Another major factor is the structure or spatial form that these molecules can assume. It should be intuitively obvious to us that a chain molecule of high molecular weight will possess a great deal of structural versatility. In the ultimate analysis it is this structural versatility, and the consequences thereof, that distinguishes macromolecular systems from monomeric ones. We shall examine these structural features in great detail as we seek the underlying basis for the properties of macromolecules.

A linear chain has already been given a schematic representation in Fig. 1, Chapter II. The symbol A can be replaced by an explicit chemical formula such as $-CH_2-$ so that a polyethylene chain is being described. A more detailed representation of this chemically simple chain can be given as shown in Fig. 1. In this representation we have precisely fixed the length of the C—C bond to its accepted value of 1.53 Å and have also fixed the valence angle at its value of 109°. Even when the normal bond lengths and angles are specified, the representation

Fig. 1 Polyethylene chain.

given in Fig. 1 is not adequate. The reason for this is that we have of necessity restricted ourselves to a structure, or conformation as it is called, that is required to lie in the plane of the paper. This is an arbitrary restriction on our part and occurs because the written word is conventionally displayed in two-dimensional space. In writing this structure we have not allowed for the possibility of rotations about the single bonds that make up the chain. Such rotations can be accomplished while the normal valence angles and bond length are maintained. This capability for internal rotation allows a chain molecule to generate a large number of different conformations. Conformations are defined as the nonidentical arrangements produced by the rotation of atoms around one or more single bonds. The conformation illustrated in Fig. 1 is thus one that has conveniently, but arbitrarily, been selected from among many.

We must now ask what we mean by internal rotation and how this operation can lead to many different structures. The fundamental principle involved here can be easily understood by studying a simple molecule such as ethane C_2H_6. This molecule contains but one rotatable bond, the carbon-carbon single bond. A representation of two conformations of ethane is given in Fig. 2. At first sight these two conformations might appear to be the same. However, close examination shows that they are indeed different. The three hydrogen atoms attached to each of the carbons are arranged differently relative to one another. In Fig. 2a each set of hydrogen atoms is directly opposite each other, as indicated by the vertical dashed lines. This is known as the eclipsed conformation. If one carbon atom is now rotated relative to the other, along the axis of the C—C bond, the relationship between the two sets of hydrogen atoms will be changed. When a rotation of 60° is made about the C—C bond the structure represented in Fig. 2b results. Here, the hydrogen atoms from each set are as far from one another as is possible. This is known as the staggered conformation. The staggered conformation will be repeated for every 120° of rotation, so that there are three such structures. However, since each of the hydrogen atoms is in exactly the same relative position for each rotation of 120° from that of Fig. 2b, identical, or equivalent, structures result.

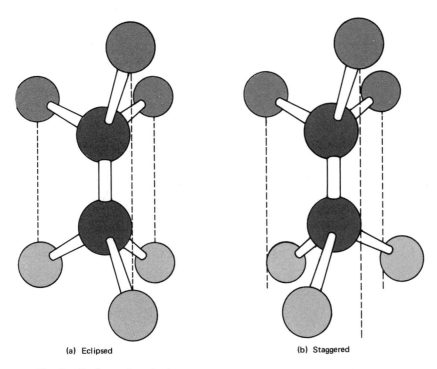

(a) Eclipsed   (b) Staggered

Fig. 2 Conformation of ethane.

From a purely operational point of view, an infinite number of conformations can be generated by rotation about the carbon-carbon bond. However, one should not conclude from this that all conformations are equally probable. The probability that a conformation will exist depends on its potential energy. Conformations characterized by low energy will be found much more frequently than those with very high energy. In the eclipsed conformation of Fig. 2a, the hydrogen atoms from each set are as close to each other as the geometric conditions allow. They are, in fact, placed so close to one another that strong repulsive forces arise. These repulsive forces make the energy of the eclipsed form much greater than that of the staggered one. Hence, it will be a very much less favored structure. Following arguments of this kind, we can represent the potential energy of ethane as a function of the rotational angle ϕ, as in Fig. 3. Each of the equivalent staggered forms are arbitrarily assigned rotational angles $0°$, $+120°$, and $-120°$, respectively. These are the conformations of lowest energy, since the hydrogen atoms are as far apart from one another as possible. The highest energy, eclipsed forms, are displaced from each of these by rotations of $60°$. The energies

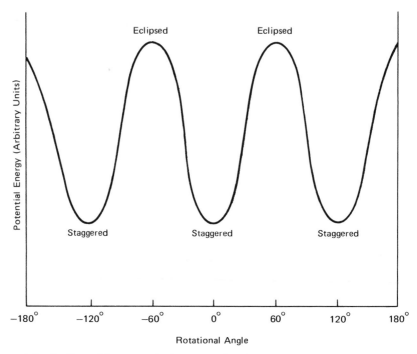

Fig. 3 Potential energy of ethane as a function of rotational angle.

characteristic of intermediate angles of rotation lie between the two extremes. We see, therefore, that although all rotational angles are geometrically possible in ethane, certain ones are preferred on energetic grounds because of the interaction between the hydrogen atoms, which are not involved in chemical bonds and are called nonbonded atoms.

The rotational freedom about a single bond is a result of many factors. A very important contribution comes from the nature of the bond itself. Most commonly found in chain molecules are the carbon-carbon, carbon-oxygen, silicon-oxygen, or carbon-nitrogen single bonds. Such bonds have an inherent rotational or torsional freedom which depends on the details of its electronic structure. Double bonds, such as are found in ethylene and in dienes, have very little rotational freedom and can be thought of as being rigid or nonrotatable. The inherent rotational characteristics of a bond will be modified by the groups that are attached to the chain atoms, called the substituents. The interactions between the nonbonded atoms will restrict the preferred rotational angles. The size of these atoms or their steric influence, is very important in this respect—as we have already seen even for the small hydrogen atoms. When bulkier substituents are attached, the rotational freedom is further restricted. The steric influences are modified further by the repulsive and attractive forces between the atoms.

A somewhat more complicated situation arises when the internal rotation in *n*-butane is analyzed. This molecule can be represented in skeletal form as shown in Fig. 4. We focus attention here on the rotation

$$H_3C-CH_2-CH_2-CH_3$$
$$1 \quad 2 \quad 3 \quad 4$$

Fig. 4 Skeletal representation of *n*-butane.

about the central bond which connects carbon atoms 2 and 3. The substituents on these two carbon atoms are no longer identical. In relation to ethane, a hydrogen atom from each of the carbon atoms has been replaced by the much bulkier —CH_3 group. The staggered conformations will clearly still be more favored than the eclipsed ones. However, in this case, not all of the three staggered conformations will be energetically equivalent. In one of the staggered conformations the two terminal methyl groups will be as far apart from one another as possible. This structure is shown in Fig. 5 and is the planar or *trans* form. It is the

Fig. 5 *Trans* conformation *n*-butane.

conformation of lowest energy. Either a clockwise or a counterclockwise rotation of 120° about the central bond will also yield a staggered conformation as is shown in Fig. 6. These two structures, which are called the *gauche* form are skewed or nonplanar. The distance between the two CH_3 groups is the same in both of these forms. These two groups are, however, closer to one another than in the planar *trans* form. In the *gauche* forms, the hydrogen atoms attached to the terminal carbons will repel one another because of the closeness of approach. Consequently, the energy of the *gauche* forms will be somewhat greater than that associated with the *trans* form. The occurrence of *gauche* conformations will, therefore, be slightly less favored. At ordinary temperatures about half as many molecules will be found in the *gauche* form as compared to the *trans*.

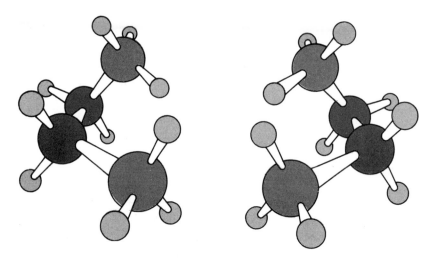

Fig. 6 The two gauche conformation of *n*-butane.

The relative energy of the *n*-butane conformations as a function of rotation about the internal carbon-carbon bond is given in Fig. 7. Here, the most energetically favored *trans* conformation is assigned the rotational angle $\phi = 0$. The two *gauche* conformations are at $\pm 120°$, respectively. Although their energies are higher than the *trans* form, the fact that they are still a favored conformation is represented by the minimum in the energy function. If we compare this diagram with the corresponding one for ethane (Fig. 3), we see that the favored conformations still occur at the same rotational angles for both molecules. However, for ethane the three favored or staggered conformations each possess the same energy.

The concept of internal rotation, as was illustrated by the discussion of *n*-ethane and *n*-butane, can now be applied to the long-chain polyethylene molecule. Let us consider a set of three successive bonds centrally located within the chain as shown in Fig. 8. The first two bonds, connecting atoms C_1 to C_2 and C_2 to C_3, form a plane that we shall take as the plane of the paper. The normal valence angle is designated by θ. Atom C_4 must be located on the indicated circle, which is the base of the cone described by the rotation of bond 3 while the valence angle is kept at θ. The exact position of atom C_4 on this circle will depend on the angle of rotation of bond 3 about bond 2. If all rotational angles were equally probable, then each position on the circle would also be equally probable. In this case, the bond is said to be freely rotating. The analysis of *n*-butane has indicated that this is not a realistic concept and will therefore not apply for a bond in the polyethylene chain. Here,

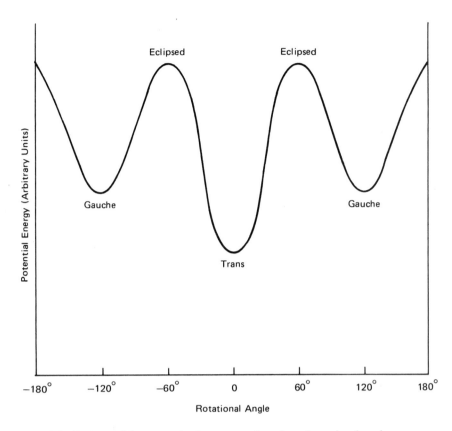

Fig. 7 Potential energy of *n*-butane as a function of rotational angle.

three preferential but not equally probable rotational angles or rotational states can be chosen. In Fig. 8, for example, position C_4' represents the *trans* conformation while C_4'' is one of the *gauche* forms. One can continue in this manner and locate the successive bond termini along the chain. The conformation, and thus the spatial form of the long-chain molecule is determined by specifying the rotational angle of each of the constituent single bonds. The molecule formed will clearly have a highly irregular form and shape and cannot be properly represented in any planar form.

Disordered Chain Conformation

For a chain of n bonds, n-2 rotational angles about the single bonds need to be specified in order to describe the conformation. However, altering the rotational position of only one bond, among the many thou-

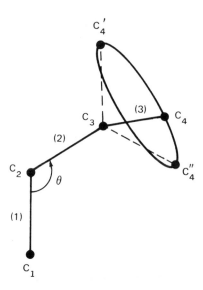

Fig. 8 Spatial representation of single bonded carbon chain. Reprinted from Paul J. Flory, PRINCIPLES OF POLYMER CHEMISTRY, Copyright 1953 by Cornell University Press. Used by permission of Cornell University Press.

sand, will change the conformation. Such alterations easily occur as a consequence of normal thermal motions. Therefore, we are faced with the problem not only of specifying a given conformation but also of enumerating all the conformations that are available to a long-chain molecule. To estimate the number of conformations that a chain molecule can assume, we will allow each bond to adopt independently one of the three preferred rotational angles. If there are n bonds per chain, then the molecule can assume 3^n distinct conformations. If $n = 10,000$, which represents a modest size chain, then 3^n is equal to $10^{4,771}$. This is a number of such astronomical proportion as to defy comprehension. As a frame of reference, it can be compared with Avogadro's number, the number of molecules in a mole. This number is a mere 10^{23} as compared to the above. It would be extremely unlikely, therefore, that two molecules would be found in the same conformation. We thus have our first insight into the tremendous structural and conformational versatility inherent to long-chain molecules. Not only is the molecular shape highly irregular but there are also an extraordinarily large number of possible conformations. The principles we have used in analyzing the structure of polyethylene will hold equally well for other kinds of long-chain molecules. Usually, only minor modifications concerned with the number, position, and energies of the preferred rotational angles are necessary.

Fig. 9 One conformation of a 1000 link polyethylene chain. Adapted from L. R. G. Treloar, *The Physics of Rubber Elasticity,* The Clarendon Press, Oxford. Reproduced by permission.

A schematic and, by necessity, two-dimensional representation of one particular structure that can be generated by the procedure outlined above is shown in Fig. 9. The arrow represents the vector distance $\vec{r}$ from one end of the chain to the other. Another pictorial method of representing a chain conformation is using molecular models as is illustrated in Fig. 10. This model represents a polyethylene conformation for a chain consisting of 150 carbon atoms. In both these models, the tortuous, random-like structure and highly irregular shape of the molecule becomes readily apparent. The molecule is highly coiled and very compact, since the end-to-end distance $\vec{r}$ is less than the extended, or contour, length of the chain. In contrast, the planar zigzag structure illustrated in Fig. 1 pg. 26 yields an extended molecule of rod-like form. However, it represents but one of the large number of conformations

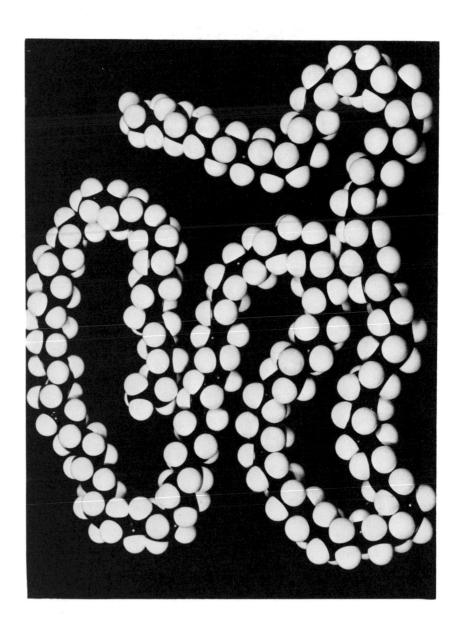

Fig. 10 Molecular model representation of a conformation of a 150 carbon polyethylene chain.

that can be generated. The overwhelming majority of these will be characterized by their compactness, irregularity, and lack of any definite order. In popular usage they have been called random coils. More properly, perhaps, they should be termed statistical conformations.

At first glance the overwhelming number of conformations available would appear to make for a rather hopeless and even ridiculous situation in which to attempt to relate the properties of macromolecules to their structure. This task would be particularly difficult if we were to use the methods employed in studying the properties of monomeric substances. For, if these procedures were adopted, we would be required to precisely locate each atom in space. To do this for all possible chain conformations would exceed the capacity of our highest speed computer. Furthermore, even if the necessary data could be assembled, they would be so numerous as to be absolutely useless for any practical purposes. Fortunately, we are concerned only with properties that can be measured. Physical measurements are the result of the average value of a property for a collection of many molecules and thus are averaged over many conformations. Hence, we need to be concerned only with the average properties of our chain molecules rather than a detailed geometric description of each. It is exactly the difficulty that we have mentioned, namely the large number of conformations involved, that makes polymer systems ideally suited to apply statistical methods in order to obtain average values.

Another question is what characteristic of a chain deserves our attention for the purpose of calculating its average value. Although there are a large number of choices, it turns out that a quantitative description of the size, or dimension, of the chain is of prime importance. A very convenient measure of the size of a chain, or its spatial characteristics, is the average of the square of r, which we shall represent as $\langle r^2 \rangle$.[1] Many properties of a chain depend on this average value. In principle this average quantity can be obtained by calculating r for each conformation, multiplying its square by the probability of the occurrence of the particular conformation, and taking the sum of this product over all the possible conformations. A direct enumeration of the desired average quantity is clearly an impossible task because of the multitudinous number of conformations involved. Fortunately, mathematical methods have been developed which avoid the necessity for direct enumeration, but still allow for an exact calculation of $\langle r^2 \rangle$ for all real chains. The only molecular information that needs to be supplied for these calculations is the number and type of bonds and the relative energies that are associated with the preferred rotational angles. The details of how

---

[1] We should recall that r represents the end-to-end distance of a chain.

these calculations are carried out are well beyond our present scope. However, we can state that the results of the exact calculations are in extraordinarily good agreement with experiment.

Despite the mathematical complexities involved in calculating the average dimensions of real chains, it is possible to adopt a simplified model that lends itself to a very straightforward analysis. A model, in general, is a hypothetical representation of the real system. Although many of the details of the model are not realistic, we can derive a very good insight into the factors that influence the dimensional properties of a chain. Furthermore, it can be shown that for sufficiently high molecular weights many of the major features of a real chain will be similar to those of the model. We take as our model a series of n identical bonds, each of length l, which are joined together in a linear sequence. The simplifications take place as we remove the requirement that valence angles between successive bonds be maintained and also allow all rotational angles to be equally probable. This chain model thus has the property that the junction of any two bonds acts as if it were a universal joint. The direction of any one bond is completely random in space and independent of the orientation and positions of any neighboring bonds. Fig. 11 shows a two-dimensional representation of such a freely jointed chain which consists of 50 bonds. Each bond is represented by a vector of equal length. The position and orientation of the bonds are uncorrelated with one another. We seek the quantity $\langle r^2 \rangle$ for a chain containing n such bonds. To obtain our results, however, we must first find the probability $P(r)$ that the two ends of the chain are separated by the distance r.

The problem of obtaining the chain dimensions is identical to the classical mathematical problem of a random walk. In the simplest form of this problem the walker takes a succession of steps of equal length which are uncorrelated with one another. In selecting the direction for a step, all memory of the path already traversed is erased. Thus, the freely jointed chain in two-dimensions, illustrated in Fig. 11, can equally well represent a random walk. The starting point or origin is at one end of the molecule, while the terminus can be represented by the other end. However, our chain, even a freely jointed one, requires a three-dimensional representation. An analogous problem is seen in the random flight of a bird where the restriction of a planar path has been removed.

For either the random walk or random flight we require the distance traversed from the origin after n steps are taken. The probability that the path will be a straight line, which will require all the steps to be in the same direction, will clearly be exceedingly small. Similarly the probability of the walker's returning to the starting point will also be very small. The exact mathematical solutions to this problem, for all

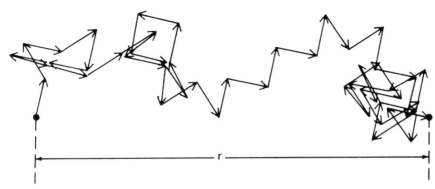

Fig. 11 Two-dimensional representation of a freely-jointed chain consisting of 50 bonds. Reprinted from Paul J. Flory, PRINCIPLES OF POLYMER CHEMISTRY, Copyright 1953 by Cornell University. Used by permission of Cornell University Press.

values of r, is given by the well-known error function of Gauss. The probability $P(r)$ is plotted graphically as a function of r in Fig. 12. The corresponding mathematical expression is given by Eq. (1).

$$P(r) = Ar^2 \exp(-b^2r^2) \tag{1}$$

In this equation A and b are constants. The constant A assures us that the total probability (for all possible values of r) is equal to unity or certainty. The constant b is related to the properties of the freely jointed chain through the relation

$$b^2 = \frac{3}{2nl^2}. \tag{2}$$

The curve in Fig. 12 has fulfilled our intuitive expectation. The probability of returning to the origin is zero. The probability of traversing very large distances is so small as to be negligible for all practical purposes. At some intermediate value of r, the probability has a maximum value. A random walk is therefore not a very efficient process by which to travel between two fixed points.

Our desired average quantity can be obtained from Eq. (1) by the standard method of averaging. It is found that

$$\langle r^2 \rangle = nl^2. \tag{3}$$

This equation tells us that the average of the square of the distance between the chain ends is proportional to the number of links in the chain. The proportionality constant is the square of the bond length. The appropriate linear dimension is the root-mean-square value of r and is expressed as

$$\langle r^2 \rangle^{1/2} = n^{1/2}l \tag{4}$$

37

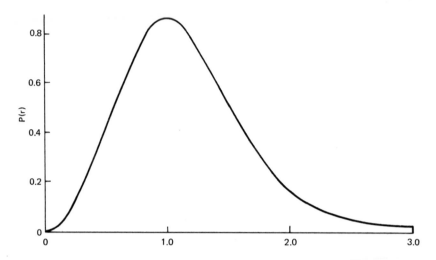

Fig. 12 Curve representing the probability of finding the two ends of a freely-jointed chain separated by a distance r. Graphical representation of Equation (1). From L. R. G. Treloar, *The Physics of Rubber Elasticity*, The Clarendon Press, Oxford. Reproduced by permission.

From Eq. (4) we find that a linear dimension of the chain will increase only in proportion to the *square root* of the number of steps. This quantitative analysis of a random walk explains why a freely jointed chain will be much more compact than an extended or stretched-out chain. In this latter conformation a linear dimension must of course be proportional to the number of bonds in the chain.

We have recognized earlier that a real chain molecule cannot be considered to be freely jointed. Therefore, it cannot be legitimately subjected to this particular random walk analysis. The influence of valence angle restrictions as well as the preference for certain rotational angles must be taken into account in describing the dimension of a real molecule. Fortunately, a detailed analysis shows that these factors can be properly and exactly taken into account merely by inserting a constant into Eq. (3). Thus, for a real chain we can write

$$\langle r^2 \rangle = Cnl^2. \tag{5}$$

The constant C will vary from one chain type to another and will depend on the chemical nature of the repeating unit, as is reflected in the bond angles and the hindrance to rotation. Units with different kinds of bonds can also be accounted for. Invariably, however, the proportionality between $\langle r^2 \rangle^{1/2}$ and $n^{1/2}$ is maintained for all real chains. For polyethylene, the constant C is found to be 6.8. Thus, for a chain where $n = 10^4$ and

$l = 1.53$ Å, we find by substituting in Eq. (5) that $\langle r^2 \rangle^{1/2} = 320$ Å. There are several experimental methods that enable us to determine the quantity $\langle r^2 \rangle^{1/2}$ directly. Very good agreement is found between the theory and direct observation. On the other hand, a fully extended chain (all *trans* rotational angles) containing the same number of bonds will be 12,700Å long. Thus, we can see quantitatively that the average end-to-end distance of a real chain, in a statistical conformation, is severely reduced relative to its extended length. If we take the value of $\langle r^2 \rangle^{1/2}$ to represent the diameter of the domain of the random coil, the volume for the above chain is 3.5×10^7 Å³. This volume is very much greater than the volume actually occupied by the constituent atoms.[2]

The influence of the chemical nature of the repeating unit on the average dimensions of chains in statistical conformation is manifested in the value of the constant C. This constant can be determined experimentally, and a compilation of typical values for different polymers are given in Table I. The values of this constant range from 2 to about 10. In all cases the structures are relatively compact. They are, however, greater than what would be expected for a freely jointed chain or even for a freely rotating chain with fixed valence angles. For a freely rotating polyethylene chain $C = 2$. The actual chain is more extended since $C = 6.8$. There are small but well-defined differences between the different chemical types. These differences have their origin in the nature of the rotatable bonds and the interactions that occur between the nonbonded atoms.

Ordered Conformations

Our discussion of chain conformations has been focused primarily on the disordered or statistical conformations. We have tactily assumed that successive rotational angles will be uncorrelated with one another. Molecular forms are thus generated which do not represent any well-defined shapes. These structures, typical of all chain molecules, represent an overwhelming preponderance of the available conformations. They indeed serve as the basis for a large number of properties. However, we must recognize that among these multitudinous conformations there

[2] In the treatment of chain molecules within the framework of random flight theory, a fundamental difference between the trajectory of a bird and the conformation of a chain has been neglected. A bird in random flight can recross its path with impunity as many times as is desired. However, since a chain element has a finite volume, intersecting paths are not allowed since two elements cannot occupy the same space. An exact mathematical analysis that eliminates intersecting conformations has not as yet been accomplished. An approximate, but very reliable solution indicates a modification of the dependence on n in Eq. (3). This important result does not, however, impose a serious constraint for our purposes.

Table I*

Characteristic Dimensions of Chain Molecules

Polymer	Repeating Unit	Temperature °C	$C = \langle r^2 \rangle / nl^2$
Polyethylene	$-CH_2-CH_2-$	140	6.8
Polypropylene (isotactic)**	CH_2-CH $\quad\ \vert$ $\quad\ CH_3$	145	5.7
Polyisobutylene	$\qquad CH_3$ $\qquad \vert$ CH_2-C- $\qquad \vert$ $\qquad CH_3$	24	6.6
Polymethyl methacrylate (isotactic)**	$\qquad CH_3$ $\qquad \vert$ CH_2-C- $\qquad \vert$ $\qquad COOCH_3$	25	9.1
Polymethyl methacrylate (syndiotactic)**	$\qquad CH_3$ $\qquad \vert$ CH_2-C- $\qquad \vert$ $\qquad COOCH_3$	35	7.2
Polyoxyethylene	$O-CH_2-CH_2$	35	4.0
Polydimethyl siloxane	$\qquad CH_3$ $\qquad \vert$ $-O-SI-$ $\qquad \vert$ $\qquad CH_3$	20	6.2
Polyhexamethylene adipamide	$-NH(CH_2)_6-NH-$ $-CO(CH_2)_4CO$	25	5.9
Poly α amino acids	$-NH-CH-CO-$ $\qquad \vert$ $\qquad CH_2R'$	25–100	8.5–9.5
Polyglycine	$-NH-CH-CO$ $\qquad \vert$ $\qquad H$	25	2.0
Natural Rubber	$\qquad CH_3$ $\qquad \vert$ $CH_2-C=CH-CH_2$	50	4.7

*Adapted from P.J. Flory, Statistical Mechanics of Chain Molecules, Wiley (1969).

**The two stereoisomers of vinyl polymers will be discussed in the next section.

exist a very few distinct ones in which successive rotational angles are correlated with one another. A definite set of rotational angles will be repeated in a very prescribed manner along the chain. We have already encountered one such simple example in the extended planar zigzag form of polyethylene. Here each rotational angle assumes the *trans* position. More generally, when a given sequence of rotational angles successively repeats itself, an ordered structure will result and the molecular shape will have a rod-like character. In analogy, while a random walk leads to a disordered conformation, a directed or defined walk results in an ordered conformation. In nonmolecular terms, a comparison can be made with a long piece of spaghetti. Before being cooked, it has the form of a long thin rod. After cooking, a much more compacted, coiled up, irregular structure results which is continually changing its size and shape. The former situation can be thought of as representing an ordered structure. The latter, a statistical or random one.

Ordered structures of isolated chain molecules as would be observed, for example, in very dilute solutions have been found for polymers of biological interest such as the polypeptides and nucleic acids. We shall examine these classes of macromolecules separately in Chapter VI. For all other chain molecules, on the other hand, ordered structures are observed only under special conditions of temperature and pressure when a large number of them are brought together in the solid state. Under these circumstances, it is possible to discuss the conformations or structure of a single chain.

To form a stable, ordered chain structure the set of rotational angles that are successively repeated must be such that each angle represents an energy minimum. Thus, a very energetically favored conformation is developed. For polyethylene, this condition is achieved when each bond assumes the *trans* position, as was illustrated in Fig. 1. A planar, or nearly planar, extended ordered conformation is characteristic of many polymers, such as the polyamides and polyesters. For example, as is shown in Fig. 13, the observed ordered structure of the aromatic polyester, polyethylene terephthalate, is nearly planar. The chain length is just slightly less than that for full extension. For polyamides, in ordered conformation, the chains are in a planar zigzag conformation and are fully extended. The ordered structures of two commonly used polyamides, nylon 6-6 and nylon 6, are represented in Fig. 14. For each of these polymers, three chains arranged side by side are shown. The planar zigzag form is readily apparent for each of the individual chains. In this conformation, the geometry and atomic coordinates are such that a very favorable strong interaction takes place between the C=O group of one chain and the N—H group of the adjacent chain. Such interactions are known as hydrogen bonds. Every possible C=O—H—N hydro-

Fig. 13 Ordered chain conformation of polyethylene terephthalate. From R. Dubeny, C. W. Bunn and C. J. Brown, *Proceeding Royal Society 226A, 531* (1954). Reproduced by permission.

gen bond is formed for these chains. These exert a very strong influence on the physical properties of the nylons.

When one or more of the hydrogens from a carbon-carbon single bonded chain are replaced by other atoms or groups of atoms the planarity of the structure is very often disturbed. Substituents larger than hydrogen can cause severe steric or packing problems when the *trans* orientation of each bond is maintained. To alleviate this very unfavorable energetic situation, other rotational angles, or succession of rotational angles, must be adopted. When these deviations from the *trans* position are repeated in a systematic manner, a spiral or helical form of the chain must evolve. We can illustrate this point in a purely formal way

Fig. 14 Ordered chain conformation for polyamides. (a) Nylon 6-6; (b) Nylon 6. From D. R. Holmes, C. W. Bunn and D. J. Smith, Journal of Polymer Science *17*, 159 (1955).

by allowing a *trans-gauche* sequence to be perpetuated along the chain. As is diagrammatically represented in Fig. 15, a helical structure results. Since there are two possible *gauche* positions obtained by rotation of plus or minus 120° from the *trans* (see Fig. 6), the same *gauche* angle must always be taken to preserve the order. Depending upon which

Fig. 15 Diagramatic representation of helical structure that is generated by a succession of *trans*-gauche rotational angles. Backbone carbon atoms ⊛; substituent group ●; hydrogen ○.

of the *gauche* orientations is selected, either a right or left-handed helix will be generated. The planar zigzag form is merely a special case of a fully stretched out spiral restricted to a plane.

A simple example of the deviations from planarity is seen in the case of polytetrafluoroethylene, or Teflon. In this molecule all the hydrogen atoms of polyethylene have been replaced by fluorine. Because of the increased atomic size, albeit a very modest one, the all *trans* structure can no longer be maintained. As is shown in Fig. 16, the chains have

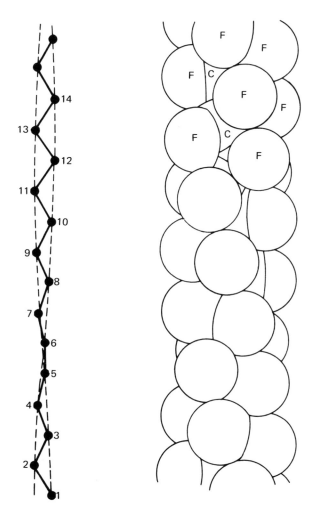

Fig. 16 Ordered, twisted conformation of polytetrafluoroethylene. From C. W. Bunn and E. R. Howells. *Nature* **174,** 549 (1954). Reproduced by permission.

to be twisted out of the plane at each carbon atom by 14° to accommodate the fluorines. A spiral structure results, which repeats itself after
every 13 carbon atoms. As is indicated in the figure, carbon atoms labeled
1 and 14 are thus in equivalent spatial positions.

Another molecule of interest is polyisobutylene. Here the pairs of

$$\left(\begin{array}{c} H \;\; CH_3 \\ |\;\;\; | \\ C\!-\!C \\ |\;\;\; | \\ H \;\; CH_3 \end{array} \right)_n$$

CH_3 groups on the alternate chain carbon atoms cause a very severe
overcrowding. These steric difficulties cannot be overcome by any combinations of bond rotations which are restricted to the *trans* or *gauche*
states. A stable, ordered, helical structure can be formed with this chain
by rotating each bond 82° from its *trans* state. This rotation generates
a helix in which eight chemical repeating units correspond to five turns
of the geometric repeating unit. An ordered structure will result when
the direction of the rotation is the same for each bond. However, when
the direction of the rotation is allowed to change at alternate bonds
a disordered statistical conformation will result. The ordered conformation of polyisobutylene was one of the first helical structures discovered
in long-chain molecules.

Vinyl polymers of the type $\left(\begin{array}{c} H \;\; H \\ |\;\; | \\ C\!-\!C \\ |\;\; | \\ H \;\; R \end{array} \right)_n$, in which dissimilar sub-

stituents are attached to alternate carbon atoms, present a very special
situation since two different structures are possible. This situation is
illustrated in Fig. 17 for a repeating unit selected from within the chain.
In this figure the carbon-carbon bond, which forms the chain backbone
lies in the plane of the paper. Attached to one of the carbons are the
two hydrogen atoms, one directed above the plane (solid line) and the
other directed below (dashed line). These two hydrogens are equivalent
to one another and are thus the same for each of the structures. However,
for the other carbon atom in which two chemically different substituents
are attached, the situation is not the same. In Fig. 17a the CH_3 group
lies above the plane, while the hydrogen atom is below. In Fig. 17b,
however, the opposite situation exists. The hydrogen atom is directed
above the plane, while the CH_3 is directed below. Neither form can
be converted into the other by any possible internal rotations about
the C—C bonds or any symmetry operation involving the whole chain.
They are distinctly different chemical structures called stereo-isomers,

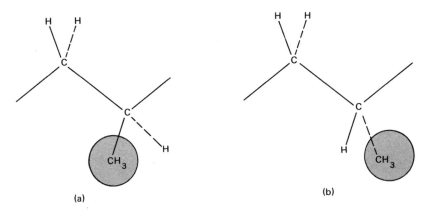

Fig. 17 Alternate arrangements of substituents about carbon-carbon bond of polypropylene.

which are mirror images of one another. They are related to each other in the same way as the left hand is to the right. A chain comprised of both of these units is a stereo-irregular copolymer, and is given the name atactic.[3] When only one of the stereo-isomers is perpetuated along the chain a stereo-regular polymer called isotactic is formed. Stereo-regular polymers are also formed when there is a regular alternation between the two isomers along the chain. Such a chain has been named syndiotactic. Chains with a high degree of stereo-regularity were first synthesized in the 1950's by Professor G. Natta of Milan. The polymers were prepared by ionic addition polymerization methods using a stereo-specific catalyst developed by Professor K. Ziegler of Germany. For this outstanding accomplishment in synthetic chemistry Natta and Ziegler received the Nobel Prize in 1963.

The stereo-regular forms of such chains can be schematically represented as in Fig. 18a and b. The atactic form is shown in Fig. 18c. We take the specific case of polypropylene, R = CH_3, and to illustrate the stereo-isomers take the carbon-carbon backbone to be planar zigzag. In the isotactic structure all the CH_3 groups are positioned above the plane, (or all below); while in the syndiotactic form there is a regular alternation of the CH_3 groups above and below the plane. In the atactic form the CH_3 groups are randomly positioned along the chain. If, however, we examine the planar zigzag form of isotactic polypropylene more closely, we find the distance between successive CH_3 groups is only 2.5 to 2.6 Å. This small separation cannot be tolerated since it leads to gross overcrowding of the methyl groups. This severe steric problem

[3] Tactic comes from the Greek word meaning order.

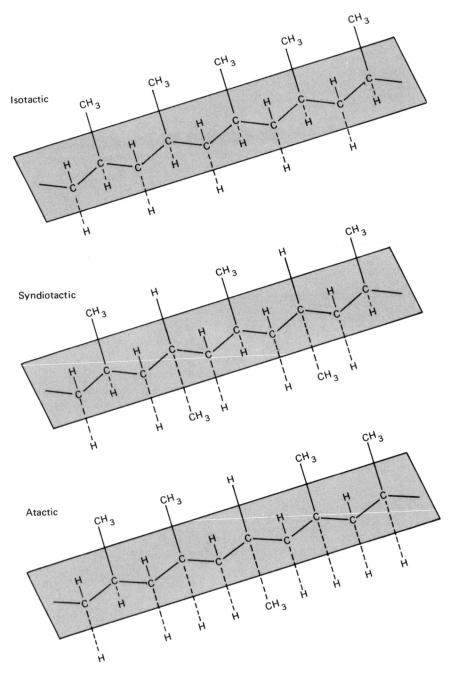

Fig. 18 Schematic representation of stereoisomeric forms of polypropylene.

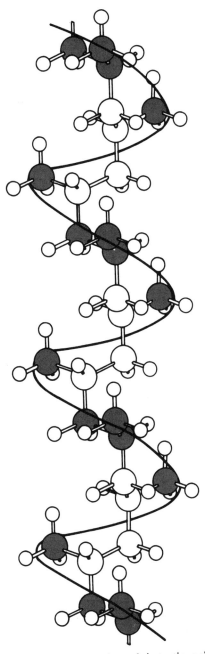

Fig. 19 Ordered helical conformation of isotactic polypropylene. Adapted from F. Danusso, Polymer *8,* 281 (1967).

can be overcome very simply by having alternate chain bonds assume a *gauche* orientation. As we have discussed the planarity of the chain is now lost and a simple helical form evolves. In this new geometrical pattern the CH_3 groups are now 3.2 Å apart. This is a sufficient separation to eliminate the unfavorable steric interaction. As is shown in Fig. 19, the helical structure particular to polypropylene contains three chemical repeating units for each repeat of the geometric structure. Isotactic polymers which contain bulkier side groups will naturally require more space. Hence helices with larger repeats will be generated. For syndiotactic polymers the substituents will be further apart as the diagram of Fig. 18b indicates. The steric problem between neighbors is not as severe as for isotactic polymers. Planar zigzag ordered structures are therefore still possible and are in fact observed in many cases. An example of this is found in syndiotactic polyvinyl chloride.

In this chapter we have summarized the conformational or structural properties of long-chain molecules. There are two major conformational categories. The disordered or statistical conformations encompass most of the allowed possibilities. However, there also exist very important and unique highly ordered chain structures. There is a very distinct similarity and analogy here between the liquid and crystalline (solid) states of monomeric substances. The liquid state is characterized by the disorderly, random arrangement of the molecules. The crystalline state is represented by a precise and orderly molecular arrangement. We should, therefore, expect that when a collection of macromolecules is brought together to form a macroscopic or observable system the resultant properties will depend on the conformation of the individual chains. Put another way, the chemical and physical properties will depend upon whether or not the chains are in disordered conformations. In the following chapters, we examine in more detail the major properties of polymeric systems and their relation to chain structure.

Chapter IV

RUBBERS AND GLASSES

Rubbers—Introduction

The unusual elastic and mechanical properties of rubber and rubber-like materials are well known to us from everyday experience. Rubbers represent a unique class of substances since they can undergo very large deformations without breaking and spontaneously return to their original dimensions when the applied force is removed. A typical rubber can be reversibly extended from five to ten times its original length. Such a deformation is called an elastic one. In contrast, ordinary monomeric solids, such as metals and crystals, can only be elastically deformed about 1 percent. Further deformation of these materials results either in rupture or in plastic flow. Plastic flow is an inelastic deformation, for when the applied force is removed the sample remains in the deformed state. Thus, the extensibility of a rubber is about 1,000 times greater than an ordinary solid. Besides the differences in extensibility, the forces required to achieve the same deformation for these two classes of substances are very different. For example, if a steel wire and rubber filament of the same diameter were stretched only 1 percent of their original length a 10^5 times greater force would be required for the steel wire. In Fig. 1 we compare typical force-length curves of rubbers and monomeric solids. The monomeric material is relatively inelastic since a large force is required to achieve a small extension. On the other hand, five- to sixfold increases in length are easily achieved with the rubber by the application of only modest external forces.

Long-range elasticity is a property found only in polymeric substances. However, it is not restricted to chain molecules with a particular kind of chemical repeating unit commonly associated with rubbers or elastomers. It is in fact observed for all types of long-chain molecules when they exist in statistical conformation, i.e., in the disordered state. Under appropriate conditions of temperature and composition, identical long-range elastic properties are observed in polyesters and polyamides in denatured globular and fibrous proteins, and in the protein elastin found in elastic tissue and in the walls of blood vessels. We must conclude

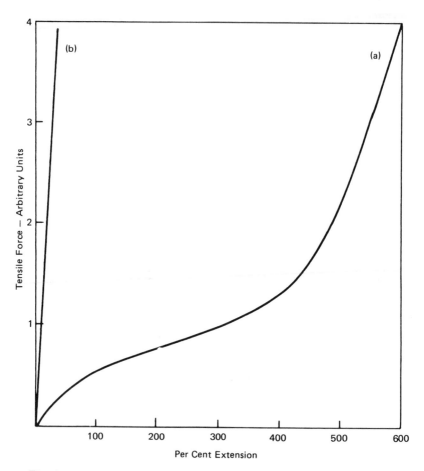

Fig. 1 Schematic representation of force-extension curve for (a) a rubber-like substance; (b) a monomeric solid.

from the diversity of chemical types that display this elastic behavior that it is the result of a common underlying structural feature unique to long-chain molecules.

Before examining the structural basis for the long-range elasticity, we should consider some associated properties. It is well known that nonpolymeric substances will expand on heating, whether they are in an unstressed or stressed condition. This normal thermal expansion behavior is obeyed by virtually all monomeric substances.[1] Polymeric materials, when undeformed, follow this classical pattern. However, Gough discovered almost two centuries ago that when a stretched piece of rubber

[1] An obvious exception to this generalization is the behavior of water at 4°C.

is heated, its length diminishes. These two kinds of thermal expansion are illustrated in Fig. 2. Stretched rubbers thus have a negative thermal expansion coefficient; contraction is observed in heating and expansion on cooling. This behavior is found in all deformed polymeric systems when in the disordered state.

Closely related to the negative thermal expansion coefficient is the observation that when a rubber sample is extended rapidly its temperature rises. This phenomenon can be demonstrated very simply by rapidly

Thermal Expansion of a Rubber-Like Substance

(a) Unstressed

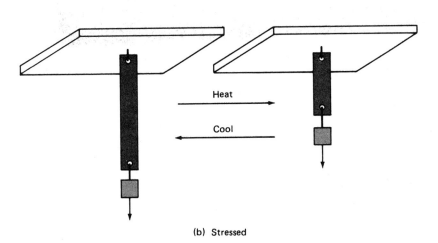

(b) Stressed

Fig. 2 Thermal expansion behavior of a rubber-like substance (a) unstressed; (b) stressed.

extending a rubber band and immediately placing it on one's lips. The change in temperature is easily detected. More sophisticated measurements show that there is about a 10°C temperature rise for a four- to fivefold extension. A student of thermodynamics will recognize that these rather special thermal properties are directly related to one another. From the Second Law of thermodynamics one can conclude that the deformation of a rubber-like substance must be accompanied by a decrease in the entropy.

A direct connection between these unusual mechanical and thermal properties can be made through a consideration of the molecular structure of long-chain polymers in the disordered state. We have already noted that macromolecules in the unstrained state possess a tremendous conformational versatility as a result of the rotational freedom about the single bonds comprising the chain. A variety of structures, ranging from very compact ones to more extended ones, are thus available. When subject to an external force, the conformations will be rearranged. In particular, the more extended forms will now be ones that will be favored. A polymeric substance can, therefore, sustain large deformations without rupture because molecular structures that are consistent with the increase in macroscopic dimensions are available. The deformation capability is thus a reflection of the change in molecular conformation without requiring any change in the intrinsic molecular arrangements. When deformed, the number of conformations available must, however, be reduced. A decrease in the entropy results, as is shown by the thermal properties that are observed.

Since the original, undeformed dimensions of a rubber-like material can be easily regained, a large restoring force must have developed. This restoring force is a reflection of the strong desire of the chains to return to their original statistical conformation. Put another way, the force causing a rubber-like substance to snap back is merely the manifestation of the desire to return to the most probable state. In this state all possible conformations, unbiased by the external stress, can be assumed. The most probable state, that of the maximum entropy, is desired by all molecular systems. It is, therefore, not a special polymeric property. Thus, the unique property of long-range elasticity is directly and intimately associated with the characteristic and novel conformational properties of polymers. In contrast, the elastic deformation of monomeric solids and liquids results in the displacement of individual atoms relative to one another. The molecular mechanism for these substances is quite different from that for polymers. It is, therefore, not surprising that for monomeric materials the maximum deformation that can be attained and the thermal properties are also different.

The fact that rubber elasticity is based on simple, well-established physical laws can also be seen in the conceptual similarity between it and the compression of an ideal gas. When the volume occupied by a gas is reduced by increasing the pressure, then the number of configurations or arrangements of the gas molecules is also reduced. The entropy of the system is decreased, and heat is evolved. As we have already noted, when a rubber band is stretched, the entropy is also decreased and heat is evolved. The pressure exerted by an ideal gas, at constant volume, is directly proportional to the absolute temperature. As we shall see, the force exerted by a rubber band, held at fixed length, is also directly proportional to the absolute temperature.

Structural Basis for Rubber Elasticity

The molecular concepts underlying long-range elasticity can be given a more quantitative description. The formal methods of statistical mechanics are used to express the relationship between the number of conformations available to the system, the entropy, and the state of deformation. By following these procedures, the force-length relationship can be derived. As a first step in this process it is instructive to develop an expression for the tension exerted on a single isolated long-chain molecule. This is clearly a hypothetical situation, since one cannot grab the two ends of an isolated molecule and move them apart. However, although we will ultimately be concerned with the macroscopic properties of a collection of such molecules, an understanding of the elastic properties of one such molecule provides the basis for understanding the macroscopic system.

For present purposes we shall take a freely jointed chain to represent our real molecule. The mean-square end-to-end distance, $\langle r^2 \rangle^{1/2}$, is then given by Eq. 4 of Chapter III. The probability of finding a distance r between the two ends of the chain is given by Eq. 1 of Chapter III and is illustrated in Fig. 12 of that chapter. We now imagine such a molecule with its ends separated by a distance r and held fixed at points A and B as in Fig. 3. If the ends were not fixed, then the end-to-end-distance would be continually changing as the molecular conformation rearranges as a result of the thermal motion of the chain elements. When the chain ends are fixed, as in the deformed condition, the motion of the chain elements acts to reduce the distance r to its most probable value. This action generates a force directed along the line AB. Therefore, to maintain r at a fixed value, an oppositely directed external force must be applied to the chain ends. This force is indicated by the symbol f in the figure. The origin of the tensile force is thus very analogous to the pressure developed on walls of a container by a gas molecule constrained to a fixed volume. By applying statistical principles

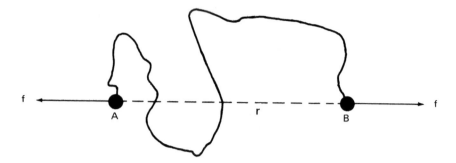

Fig. 3 The force on an isolated chain whose ends are held fixed. Adapted from L. R. G. Treloar, *The Physics of Rubber Elasticity,* The Clarendon Press, Oxford. Reproduced by permission.

it is found that the tensile force can be expressed as

$$f = 2kTb^2r \tag{1}$$

where k is Boltzmann's constant, T the absolute temperature, and b is the same constant that appeared in Eq. 1 of Chapter III. We find, therefore, that the force is directly proportional to the absolute temperature. It acts in the direction of the line joining the chain ends and is directly proportional to the distance between them. Thus, a deformed, isolated chain molecule obeys Hooke's Law, since the force is directly proportional to the displacement.

It now remains for us to relate the elastic behavior of an individual chain to that of a real macroscopic sample. Such a sample consists of a collection of a large number of chain molecules. We must first of all make sure that upon deformation the molecules will not slip or flow past one another. The flow will, of course, be much less than would occur in a monomeric liquid because of the entanglement and interweaving of the long chains with one another. However, it is still of sufficient intensity so that a pure elastic deformation will not result. Fortunately, this problem can be easily solved by introducing intermolecular cross-links to form a three-dimensional network structure. The number of cross-links required is not very large: about one-tenth of 1 percent of the chain repeating units is usually sufficient. This number is adequate to prevent flow, but sufficiently small so as not to retard the segmental motion necessary for the conformational versatility. In rubber technology this cross-linking process is known as vulcanization. Charles Goodyear's discovery, in 1839, that natural rubber can be cross-linked with sulfur when the mixture is heated paved the way for the practical and commercial use of rubber. Although the fine details of the chemistry involved in Goodyear's vulcanization process are still not completely understood, the double bonds of the isoprene repeating unit are clearly involved. Schematically, the cross-linking between two isoprene chains can be represented as shown in Fig. 4. For this type of reaction, where units from

$$--HC-\underset{\underset{S_x}{|}}{\overset{\overset{CH_3}{|}}{C}}-CH=CH-CH_2--$$

$$--CH_2-\underset{\underset{CH_3}{|}}{C}-CH_2-CH_2-CH_2--$$

Fig. 4   Natural rubber chains cross-linked with sulfur.

different chains are joined in pairs, four chains will emanate from each cross-link in the network. When natural rubber is excessively cross-linked, however, its elastic properties are lost. It becomes a hard, rigid material known as ebonite.

A schematic representation of network formation followed by deformation is given in Fig. 5. One starts with a collection of undeformed randomly coiled chains (Fig. 5a). Intermolecular cross-links are then introduced to form a three-dimensional network. A chain can now be defined as that portion of the structure that extends from one cross-linkage to a successive one along the contour of a given molecule. Thus, for example, the sequence of units traversing the contour from A to B in Fig. 5b represents one chain, whose vectorial end-to-end distance is represented by $\bar{r}$. Our network will be characterized by ν such chains, ν being the number of cross-linked units. Upon stretching, as is shown in Fig. 5c, the network becomes distorted. The chain contours naturally become more extended. The set of vectors that characterizes the end-to-end distances of the chains is considerably altered in both direction and magnitude. The vectors will become strongly biased in the stretching direction, as can be seen from following the trajectory of the set of typical vectors from Fig. 5b to Fig. 5c.

Our problem now is to relate the macroscopic dimensions of the deformed sample to those of the individual chains making up the network. This is done by recognizing that the cross-linkages represent fixed points in the network structure. However, when a network is deformed,

(a) Collection of
Linear Chains

(b) Three-Dimensional
Network

Crosslinking

Deformation

Stretching
f ← Direction

A

B

Stretching
Direction → f

(c) Deformed Network

Network Formation and Deformation

Fig. 5 Schematic representation of network formation and deformation. Reprinted from Paul J. Flory, PRINCIPLES OF POLYMER CHEMISTRY, Copyright 1953 by Cornell University. Used by permission of Cornell University Press.

these junction points must take on a new set of average spatial positions relative to one another because of the distortions imposed by the emanating chains. The coordinates describing their position in space must be altered (relative to the undeformed state) in proportion to the dimensional changes of the bulk sample. Put another way, the alteration in the length and direction of the chain vectors must correspond to the macroscopic changes. This requirement is a natural consequence of the isotropic character of the network.

When the foregoing connection is made between the microscopic (molecular) and macroscopic dimensional properties, it is found that in simple extension the stress, corresponding to a length L, can be expressed as

$$\tau = NkT \left[\frac{L}{L_o} - \left(\frac{L_o}{L} \right)^2 \right], \tag{2}$$

where τ is the force per initial cross-sectional area, L_o is the initial length, and N is the number of chains per unit volume (ν/V).[2] The theoretical result embodied in Eq. (2) points out that Hooke's law will not be obeyed by a bulk rubber sample. As will be recalled, this is contrary to the results obtained for an isolated chain.

In an exhaustive set of experimental data, involving a variety of rubbers of different chemical types, the force at fixed length, is directly proportional to the absolute temperature. This is one of the important conclusions drawn from Eq. (2). The factor in front of the brackets in this equation is the elastic modulus. It can thus be directly predicted from molecular theory. Within the uncertainty with which the number of cross-links can be determined by chemical analysis, reasonably good agreement is found between the calculated and observed modulus. It is not very often that such a physical property can be predicted theoretically.

The force-length relation expected from Eq. (2) is compared with the actual observed experimental results in Fig. 6. The theory and experiment are in very good accord for small extension ratios, (L/L_o). Semiquantitative agreement is obtained for extension ratios up to about five. At very large deformations there are indeed some major discrepancies. These can be attributed to other phenomena that are not explicitly taken into account in the simple formulation leading to Eq. (2). Although complete quantitative agreement between theory and experiment has not as yet been attained, a major advance has been made in relating molecular structure to physical properties. The long-range elasticity of rubber-like materials can thus be put on a quantitative molecular basis. Throughout the analysis the conventional laws and methods of physical science have been used. It was crucial in this development, however, that the underlying structural features of polymer molecules be recognized. This endeavor represents a classical achievement and demonstrates the great unity of concept that exists in natural science. We must, however, recall with some humility that more than a century elapsed between the observation of the unique thermoelastic properties of rubber and a molecular understanding of its elastic properties.

Glass Formation

We have noted that the existence of long-chain molecules, in disordered conformations, is a necessary condition for the development of long-range elasticity. This is not, however, a sufficient condition. The

---

[2] These calculations are not limited to simple extension or stretching—they can be carried out for other types of deformation as well.

Fig. 6 Comparison between experiment and theory of the force-length relation of stretched natural rubber. Solid line is the theoretical expectation. Open circle the experimental points. Adapted from L. R. G. Treloar, *The Physics of Rubber Elasticity,* The Clarendon Press, Oxford. Reproduced by permission.

tacit assumption was made that when an external force is applied, the time necessary for conformational adjustments to take place is very small in comparison with the time scale of the observation, i.e., the time in which the measurements are made. It has, therefore, been implicit in our discussion, heretofore, that the chains were endowed with sufficient internal or segmental mobility to allow the required conformational rearrangements and adjustments to take place. However, it is quite possible that when the temperature is lowered or the pressure raised, the time required for these changes will become comparable with or even greatly exceed the observation time. Under these conditions, despite the fact that the chains are in random conformation and the system is disordered, rubber elastic behavior will no longer be displayed. To be rubber-like, the chains must be able to sample all the available conformations. When this condition is not fulfilled, instead of a flexible, tough, and highly deformable material one obtains a material that is rigid, very brittle, and easily fractured. The energy imparted by an external force can no longer be dissipated by dimensional changes. Consequently, fracture results. This state is called the glassy state. Although glasses and rubbers clearly represent opposite extremes in mechanical properties, in both cases the chains are in a statistical conformation. Thus, the differences in the properties of the two states do not lie in any fundamental difference in structure or molecular arrangement, but rather in the time scale of the segmental motions and their response to external stress.

Common examples of polymeric glasses are polymethyl methacrylate (Plexiglas) and polystyrene. (Many of our childhood toys were made from polystyrene.) Their fragility and short life span are due to the fact that they are glasses at room temperature. Natural rubber displays all the expected elastic properties at room temperature. However, when cooled to liquid nitrogen temperature, its long-range elastic properties are lost and it becomes very brittle. This change in properties can be dramatically illustrated by taking an ordinary rubber ball and immersing it for a few minutes in a vessel of liquid nitrogen. When the ball is removed from the vessel and thrown against the floor or wall, instead of rebounding in the usual manner it will shatter. At the very low temperature of liquid nitrogen it has become a glass and consequently lost its rubber-like properties. Glass formation is not, however, restricted to polymeric substances. Monomeric liquids, such as sugars and alcohols, that do not readily crystallize also can form glasses with physical properties similar to those of polymeric glasses.[3] Many sweets—lollipops, for example—are technically glasses and must be handled carefully, lest they fracture.

The transition from a rubber to a glass is reversible—a substance can be transformed from one state to the other by appropriate changes in the temperature or pressure. In addition to the drastic changes that take place in the mechanical properties, other properties are also altered. The reason is the same, however, in that it takes time for the internal molecular structure to readjust itself to an external change. As the two states are traversed there are discontinuities in the thermal expansion coefficient, in the compressibility, and in the specific heat. In Fig. 7 we illustrate the relation between the specific volume (reciprocal of the density) and the temperature as the rubber to glassy state is traversed for two undeformed polymers, natural rubber and polystyrene. Since they are not subject to deformation, both polymers display normal thermal expansion coefficients. However, for each polymer the slope of the volume-temperature curve changes at a well-defined temperature. At

[3] For monomeric liquids it is the time scale of the molecular rearrangements relative to the observation time that causes glass formation.

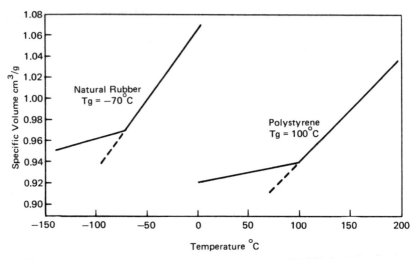

Fig. 7 Specific volume-temperature relations for polystyrene and natural rubber.

this temperature, called the glass temperature, Tg, the specific volumes are continuous but there is a discontinuity or break in the slopes of volume-temperature lines, i.e., in the thermal expansion coefficients. The thermal expansion coefficient in the rubbery state is usually two to three times greater than in the glassy state because of the greater molecular mobility. If we could carry out each observation below Tg for an exceedingly long time then we would not expect to see this discontinuity. Instead, the dashed lines would be observed. However, over the time intervals that man can be expected to make observations specific volume measurements serve as both a very reliable and convenient method of locating the glass temperature.

All polymers experience very similar changes in properties when they undergo the glass to rubber transition. The only difference is the temperature at which these changes occur. As a further example, we examine in Fig. 8 the modulus of elasticity for natural rubber and polystyrene as a function of temperature. The modulus is represented on a logarithmic scale in order to plot on one graph the large range in values. The transition region, which encompasses a 10- to 20-degree range, is characterized by about a 1,000-fold change in the modulus. The modulus is of course a reflection of the extensibility of the material. We should note that the transition temperature for these two polymers, as determined by the large change in a physical property, is the same as is found from the specific volume measurements shown in Fig. 7. The shapes of the two curves in Fig. 8 are virtually the same when cognizance

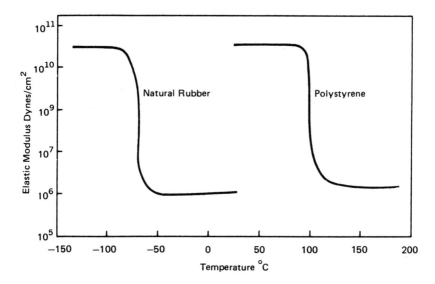

Fig. 8 Elastic modulus of polystyrene and natural rubber as a function of temperature.

is taken of the difference in glass temperatures. Below the glass temperature the modulus of elasticity is 10^{10} dynes per cm^2 for both polymers. It remains essentially constant with increasing temperature until the transition temperature. At this temperature there is a precipitous decrease in the modulus as the sample becomes a rubber. The modulus then remains essentially constant at about 10^7 dynes per cm^2 in the rubbery state. Fig. 8 illustrates quite vividly the consequences of the dual personality of a polymer chain even though it is in a statistical conformation.

Chemical Structure and Glass Formation

Table I sets forth the values of Tg for some representative polymers.

There is a wide variation in the values of the glass temperatures for organic polymers: from $-120°$C for polydimethyl siloxane to more than $100°$C for poly-α-methylstyrene. The nature of the chemical repeating unit plays a very important role in determining the temperature at which the transition between the rubbery to glassy state occurs. The major chemical factors appear to be the flexibility of the chains (i.e., the relative freedom of rotation about the single bonds), the bulkiness of side groups, and the intermolecular interactions between substituent groups. The more flexible the chain, the lower the glass temperature. Intense polar interactions and bulky side groups tend to raise this tem-

Table I
Glass Temperatures, Tg, of Polymers

Polymer	Tg°C
Polydimethylsiloxane—silicone rubber	−120
Polybutadiene	−85
Polyisobutylene—butyl rubber	−70
Poly *cis*—1,4—isoprene—natural rubber	−70
Polychloroprene—Neoprene	−50
Polyvinyl chloride	+80
Polymethyl methacrylate—Plexiglas	+100
Polystyrene	+100
Poly-α-methylstyrene	+130
Polyethylene terephthalate—Dacron, Mylar	+60
Polyhexamethylene adipamide—6,6—Nylon	+54
Polypropylene	−20

perature. The first five polymers listed in Table I are usually rubbers or elastomers at room temperature. Their rubber-like properties clearly result from the fact that their glass temperatures are located well below their use temperatures. Otherwise, the necessary response of the chain conformation to external stress would not occur and the long-range elastic properties would be lost. For the same reason polymers with high glass temperatures are not usually suitable as elastomers. However, if used above their respective Tg's they would display the usual rubber-like behavior. Such common plastic materials as polymethyl methacrylate, polystyrene, and polyvinyl chloride have glass temperatures well above their use temperature.

Plasticizers

The glass temperature of a polymeric system must naturally be altered by the incorporation of structural or molecular features that influence chain mobility. Copolymerization will effect the glass temperature, as will the addition of nonvolatile monomeric liquids. Such liquids are called plasticizers. They facilitate segmental motions and therefore act as internal lubricants. An important practical application of this principle can be illustrated in the uses of polyvinyl chloride. According to Table I, the glass temperature for the pure, undiluted polymer is 80°C. Consequently, at room temperature it is a brittle, inelastic material. Its flexibility and elastic properties can be enhanced, however, by the addition of a plasticizer that lowers the glass temperature below room temperature. Certain desirable physical properties are then given to the polymer. It has become, for example, the major constituent of plastic raincoats and garden hoses and has found extensive use in this type of application. There are, however, certain disadvantages in this use. Although nonvolatile plasticizers are used, the liquid molecules will slowly migrate or diffuse to the surface. As a result, the enhanced segmental mobility will be lost and the article will gradually stiffen and become brittle. After a while the item will no longer be useful for the purpose designed—we have all experienced the stiffening and cracking of plastic raincoats as well as the failure of garden hoses.

In this chapter we have discussed the major properties of polymers in the disordered state in which the chains have adopted a statistical or random conformation. The property of long-range elasticity, unique to long-chain molecules irrespective of their chemical constitution, can be explained in a straightforward manner when the conformational characteristics of the chains are recognized. However, we have seen that there is also a characteristic temperature at which the rubber-like properties are lost and the same polymer becomes a brittle glass. The chemical nature of the chain repeating unit, which determines the glass temperature, ultimately determines the end-use of the polymeric material. As we have noted in Chapter III, polymer chains can also exist in highly ordered conformations. The properties characteristic of this state are quite different from those we have discussed and form the subject of the next chapter.

Chapter V

CRYSTALLINE POLYMERS AND FIBERS

Crystal Structures

In this chapter we shall examine the properties of a collection of polymer molecules in the crystalline state. In this state each chain adopts the same ordered conformation. Because of the high degree of molecular order the structure in this state is very similar to that of monomeric solids. It is virtually axiomatic, and has been amply demonstrated, that polymers with a sufficiently regular chain structure will crystallize. The type of chain irregularities that retard or even suppress the crystallization process are chemically different units such as copolymer units, branch points and chemically identical units which are geometric or stereo-isomers. It should not surprise us that the drastic difference in the chain conformation is reflected in some major differences in properties as compared with the disordered state. For example, the modulus of elasticity increases anywhere from 1,000- to 10,000-fold, reflecting the loss of long-range elasticity in the ordered state.

In Chapter III we found that when an isolated long-chain molecule adopts an ordered conformation the geometric form that evolves is rod-like and asymetric. The crystalline state which is a collection of molecules in such conformations represents, in addition, order in three dimensions. Thus, each chain is not only conformationally ordered but the molecular axes are also parallel to one another. The axes are so arranged that the chain substituents form repetitive, regular planes in directions normal to the chain axis. A structural description of the crystalline state requires not only a very definite chain conformation but also a well-defined mode of chain packing. The necessary information about the molecular arrangements in polymer crystals is obtained from X-ray diffraction measurements. The experimental techniques and theoretical analysis are the same as are used in the structural determination of crystals of monomeric substances. The molecular arrangement can be described by a unit cell. This is the structural unit which is repeated in all directions and thus produces a three-dimensional lattice. Some typical ordered chain structures, all of which represent conformations of minimum energy, have been described in Chapter III. We shall show some three-dimensional structures in order to emphasize some of the packing characteristics. Fig. 1 illustrates the crystal structure of linear

Fig. 1 The crystal structure of polyethylene. From C. W. Bunn, *Chemical Crystallography*, Oxford University Press. Reproduced with permission.

polyethylene. The planar zigzag chain conformation is readily apparent. The repeating distance along the chain axis, which is 2.53 Å, represents the length of one zigzag (two carbon atoms). The unit cell is a rhombus as is indicated in the figure. The very specific regular disposition of the chains can also be seen. The cross-sectional area of the unit cell is 36 Å². The two chains per unit cell are thus positioned fairly far apart from one another.

A more complex ordered array is found for crystalline natural rubber illustrated in Fig. 2. The regularity of the chain structure is again evident.

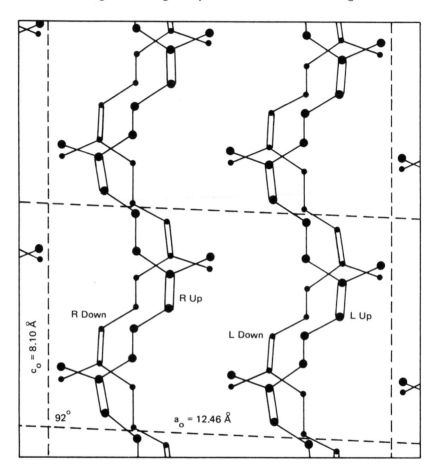

Fig. 2 The chain arrangement in natural rubber crystals. From C. W. Bunn, Proceedings of the Royal Society 180A, 40 (1942). Reproduced with permission.

The *cis* configuration is maintained repetitively about the double bonds. The repeat distance along the chain axis is 8.1 Å which indicates that there are two isoprene units per geometric repeat. Here, the chain is not in an extended conformation. The unit cell is monoclinic in this case so that the planes making up the unit cell do not meet at right angles.

The ordered chain arrangements for two polyamides have been given in Chapter III (Fig. 14). Here, the interchain distances are governed to a large extent by the strong propensity for hydrogen bonding between the O atom from the C=O group of one chain and the H atom from the N—H group of an adjacent chain. The ordered structure for another fiber forming polymer, the polyester polyethylene terephthalate, is shown in Fig. 3. The chain repeat distance is 10.75 Å, which is only slightly less than the requirements for a fully extended chain. The chains are, therefore, in a nearly planar conformation. In the unit cell the edges are of unequal length and the angles are different. There is one chain in the monoclinic unit cell.

The structures illustrated above were selected from among many examples to demonstrate the principle that long-chain molecules can be organized into a well defined three-dimensional ordered array.

Crystallization Process

At this point we may be puzzled about how such ordered structures can be formed from a polymer melt. In the molten state the chains are in disordered conformations with the characteristic irregular structures and a great deal of intermolecular entanglement. Despite the differences that must exist in the mechanistic nature of the crystallization of polymers as compared to monomers, formally the process is the same for both kinds of substances. The new state or phase must be initiated within the body of the parent liquid. This initiation process is called nucleation. The nuclei, or crystal embryo as they are sometimes called, subsequently grow into larger mature crystals. Both of these processes have been identified in polymer crystallization. The crystallization from most monomeric liquids takes place at very rapid rates at temperatures just slightly below the melting temperatures. The crystallization conditions are thus very close to that for equilibrium between the crystal and liquid. On the other hand, polymer crystallization must invariably be conducted at temperatures well below the melting temperature so that the process can proceed at an appreciable rate. Depending on the polymer, crystallization temperatures anywhere from 15° to 50° below the melting temperature are necessary. Consequently, polymer crystallizations take place under conditions well removed from equilibrium. This requirement introduces complications into the understanding of the crystalline state of polymers and sets it apart from crystals of monomeric substances.

The concept of polymer crystallization outlined so far is an obvious oversimplification. We have implied that all of the long-chain molecules,

Fig. 3 The chain arrangement in polyethylene terephthalate crystals. From R. P. Daubeney, C. W. Bunn and C. J. Brown, Proceedings of the Royal Society *226A*, 531 (1954). Reproduced with permission.

each of which is comprised of many thousands of chain repeating units, achieve a regular conformation over the complete molecular length. Furthermore, the molecules must be aligned in a parallel array with one another. Although this situation represents an idealized equilibrium requirement, the kinetic and mechanistic difficulties that must be overcome to achieve this condition are enormous. The difficulty that a chain would experience to even extricate itself from the highly disorganized molecularly entangled melt is easily understood. This process must take a great deal of time. Even if the separation of the chains were accomplished, the molecular ordering would be a kinetic, time-dependent process. In addition, the fact that polymer crystallization is invariably conducted at temperatures well below the melting temperature will contribute further to the nonequilibrium character of the final state. Because of these problems a compromise structure must be reached in any actual crystallization. A complete transformation from the disordered to crystalline state is rarely, if ever, achieved. Only a portion of a chain will adopt the required ordered conformation. Depending on the polymer type, molecular weight, and crystallization conditions the percentage of the material crystalline will vary from 30 to 90 percent. The failure of the level of crystallinity to always reach very high values, despite the great thermodynamic driving force urging the chains to do so, can be attributed to the intractability of a long chain, particularly when a portion of it becomes embedded in a crystallite. Entangled portions of the molecules that are trapped in the regions between crystallites will not be able to easily undergo the conformational rearrangements necessary for the deposition of an ordered structure on the surface of a crystallite.

A real polymer system is therefore polycrystalline, i.e., made up of many small crystallites. The crystallites have dimensions (in the chain direction) that are on the order of several hundred angstroms, and they are interspersed within a matrix of disordered or amorphous chain units. We should, therefore, not be surprised that the properties depend not only on the fact that the system is crystalline, but also on the amount of crystallinity and on the morphology. Morphology means the size and shape of the crystallites, the chain arrangement within crystallites, and their relative spatial arrangement with one another and the amorphous zones.

There are two major modes of polymer crystallization. One involves the application of an external force, so that the transformation takes place from a deformed or rubber elastic melt. In contrast, the crystallization could also be conducted without the application of an external force so that the transformation would take place from an undeformed melt. In this latter case, although each of the crystalline regions represents a highly ordered domain, the regions are randomly arranged relative

(a) Amorphous Natural Rubber

Collagen

(b) Crystalline Polyethylene

Polyethylene

(c) Oriented Crystalline Polymers

Natural Rubber

Fig. 4 Typical X-ray patterns of polymers. From *Crystallization of Polymers* by L. Mandelkern. Copyright 1964. Used with permission of McGraw-Hill Book Company.

to one another. The actual existence of such a structure can be demonstrated by the methods of X-ray diffraction. Some typical X-ray patterns are given in Fig. 4. In Fig. 4a the X-ray pattern is only a very diffuse halo. This kind of pattern is characteristic of a noncrystalline substance. The particular example is for natural rubber at room temperature, which is in the disordered or amorphous state. However, when a polymer is crystallized merely by cooling, with no deformation being imposed, a series of intense, sharp, concentric rings are found. An example of an X-ray pattern for this type of crystallization is given for linear polyethylene in Fig. 4b. This pattern is known as a powder diagram. It has the same characteristics as one obtained from a very fine crystalline powder of a monomeric substance. In this case the small crystallites are also randomly arranged relative to one another. From the position of the rings on the film the structure of the unit cell can be derived by conventional methods.

When a polymer is crystallized under a constant external force the morphological picture is changed. Because the crystallization is conducted from a deformed melt the crystallites are preferentially oriented in the stretching direction. The chain axes are thus directed along some macroscopic dimension in space. The characteristic X-ray diffraction pattern is also altered, as is shown in Fig. 4c, for some typical oriented crystalline polymers. These polymers include natural rubber, linear polyethylene, and the fibrous protein collagen. Instead of concentric rings, we now observe a set of discrete spots arranged in a very regular pattern. The fact that one observes spots instead of rings is indicative of crystallite orientation because the diffracting planes of the crystals must be biased relative to the X-ray beam. A quantitative analysis of such patterns yields the details of the orientation as well as the unit cell structure.

A similar kind of oriented system can also be obtained by another process. One starts in the undeformed crystalline state, and an external force of sufficient intensity is applied to deform the disordered regions. As a consequence, the connected crystalline regions are oriented in the stretching direction. In many practical cases a high state of crystallite orientation can be achieved more easily by this process, which is called drawing.

Undeformed Crystallization—Morphology and Properties

Crystallization in the undeformed state can be carried out over the complete composition range from very dilute solutions to the pure bulk polymer. Crystals grown from dilute solutions—i.e., polymer concentrations of the order of 1 percent or less—display a very characteristic morphological form. Under the electron microscope the crystals are found to be in the form of thin plates or lamellae. A typical electron micrograph of such crystals is given in Fig. 5. This morphological form is found

Fig. 5 Election micrograph of polyethylene crystals formed in dilute solution. From *Crystallization of Polymers* by L. Mandelkern. Copyright 1964. Used with permission of McGraw-Hill Book Company.

for all crystallizable homopolymers. The lamellae are about 100 Å thick and extend to several microns in the transverse directions. Given these dimensional requirements we would intuitively expect that the chains within the crystallite would be arranged so that their molecular axes lie parallel to the wide faces of the lamellae. However, standard methods of studying orientation have established rather convincingly that the chain axes are directed perpendicular to the wide faces. Since the platelet thickness is only of the order of 100 Å and since such crystals are formed from chains of very high molecular weight the crystallite thickness must represent only a very small fraction of the extended length of a typical molecule. Consequently, a chain must traverse a given crystallite many times in order to simultaneously satisfy both the dimensional and orientation requirements. Hence the platelets must consist of some type of folded polymer chains.

74

The very regular, smooth appearance of the crystals under the electron microscope led initially to the postulate that the chains were regularly folded or pleated. A regularly structured, smooth interface with hairpin-like bends was assumed. Although this feat is sterically possible for most chains, detailed physical-chemical studies of such crystals have shown that despite outward appearances the surfaces are not smooth or regularly structured. Instead, a disordered overlayer exists, as depicted in Fig. 6. The required chain orientation and lamellae structure is still

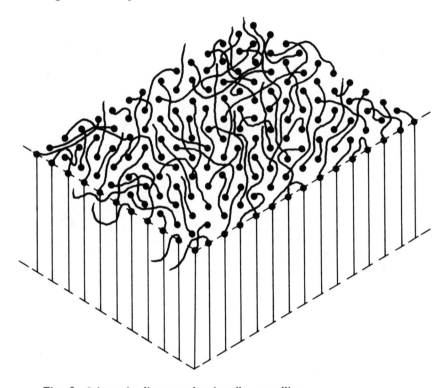

Fig. 6 Schematic diagram of a lamella crystallite.

maintained but the crystalline sequences are randomly connected by loops of random length. About 10 to 15 percent of the chain units form the disordered interfacial layer so that one does not have a completely crystalline system. This disordered layer, well established by physical measurements, cannot be resolved in the electron microscope.

The characteristic lamella morphology of crystals formed in dilute solution, with its unexpected chain orientation, is of interest not only in itself but also for its significance to the structure of polymers crystallized from the pure melt. A variety of experimental techniques, including

Fig. 7 Electron micrograph of replica of fracture surface of linear polyethylene. Scale 1 mm. = 370 Angstrom units. From Journal of Polymer Science A-2 *4*, 385 (1966). Reproduced with permission.

electron microscope observations, have established that a lamella crystallite is also the primary morphological form for bulk crystallized systems. A typical electron micrograph, showing a stack of thin lamellae with large lateral dimension, is given in Fig. 7. For bulk crystallized polymers the lamella thickness can vary from several hundred to several thousand angstroms. From a large number of detailed studies the primitive structure of a bulk crystallized polymer can be represented as in Fig. 8. The vertical lines represent the chain sequences within the crystalline regions which are in an ordered conformation. The ordered structures are either planar, zigzag, or one of the appropriate helical forms. As they emerge from the basal plane of the crystal some of the chains

Fig. 8 Schematic representation of morphology of bulk crystallized polymer.

will return to the crystallite of origin but not necessarily in juxtaposition. Others will leave the vicinity of the crystallite and will form a disordered amorphous or interzonal region which eventually joins a neighboring crystallite. The chain units in the interzonal region, which connect one crystallite with another, are in a disordered conformation. An ill-defined, disordered, and irregularly structured interfacial zone is also formed. The relative proportion of each region depends on the molecular weight and the crystallization conditions, particularly the crystallization temperature. Thus, in the crystalline state, perhaps realistically called the partially crystalline state, chain sequences which are organized into a highly ordered array and chain units in disordered conformation coexist and are connected with one another. We can therefore expect a wide range in properties as the proportions of these regions are varied for a polymer of the same chemical constitution.

Fig. 9 illustrates how some properties of linear polyethylene change with the amount of crystalline material. For this polymer the fraction of material crystalline can be varied from about 0.5 to 0.9 by control of the molecular weight and crystallization temperature. The density at room temperature ranges from about 0.99 g·cm^{-3} to 0.92 g·cm^{-3}. The former value is quite close to the density of the unit cell and thus represents an almost completely crystalline polymer. The latter value, however, represents about only half the material being crystalline. Similar changes are observed in another thermodynamic property, the heat of fusion. Here the values range from about 68 cal/gram to about 25 cal/gram. This range in values for the density and heat of fusion represent very large changes for an identically constituted chemical system. Such large variations in properties are unheard of in monomeric systems. In crystalline polymers this effect is obviously due to the two kinds of chain structures that are present and to the drastically different contribution each makes to a given property.

Besides the thermodynamic properties, changes are observed in other physical properties with the crystalline content. For example, samples with a very high level of crystallinity are very brittle and are easily fractured at room temperature. They thus can find very little practical use. As the level of crystallinity is reduced, the polymer becomes tougher and more flexible. It is no longer brittle and is not easily fractured except when cooled below its glass temperature. When the fraction of crystalline material in linear polyethylene is reduced below about 0.85 it becomes a material with a wide range of applications. The change in the elastic modulus with the level of crystallinity is also shown in Fig. 9. This quantity measures the flexibility of a material. There is about a five-fold change in the modulus as the fraction crystalline varies from 0.85 to 0.60.

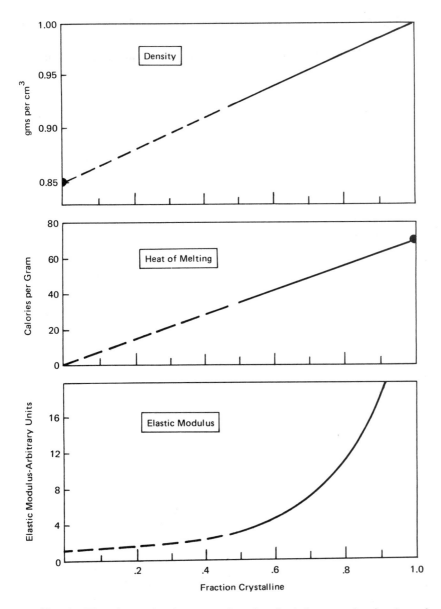

Fig. 9 The change in the properties of polyethylene as the fraction of material crystalline is varied.

Melting Process—Undeformed Crystallization

The transformation from the crystalline to the amorphous or disordered state follows a characteristic pattern for all polymers. This behavior is very similar, if not identical to the melting of monomeric solids. An example of the fusion of a high molecular weight fraction, which was crystallized very carefully and then heated very slowly, is given in Fig. 10. The specific volume, which is directly related to the fraction crystal-

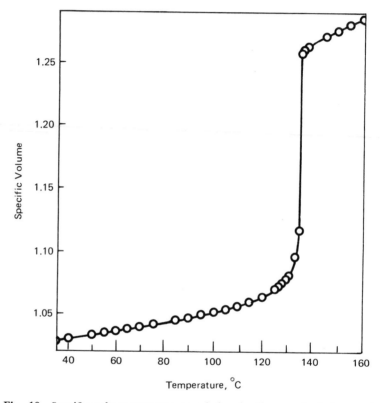

Fig. 10 Specific volume-temperature relation for linear polyethylene, molecular weight 32,000. Adapted from R. Chiang and P. J. Flory, Journal of the American Chemical Society *83*, 2857 (1961). Reproduced with permission.

line, is plotted as a function of temperature. At the lower temperatures, a linear volume-temperature relation is found which is typical of the crystalline state; at higher temperatures, the volume-temperature coefficient, although still linear, has a much higher value which is characteristic of the liquid state. About 80 percent of the transformation takes place over about a 2° temperature interval as is shown by the sharp increase in the specific volume in the vicinity of 136°C. The temperature at which the last traces of crystallinity disappear is well defined and is the melting temperature which is designated as T_m. The fusion curve illustrated is comparable in all respects to those obtained with monomeric substances. There is a characteristic discontinuity in the specific volume, or other comparable properties, at a well-defined temperature. Below the melting temperature properties of the crystalline state are observed. Above this temperature, depending on the molecular weight, the polymer will either by a viscous fluid or a rubber elastic body.

The sharpness of the fusion process illustrated in Fig. 10 and its similarity to the melting of monomers is a consequence of utilizing a molecular weight fraction and employing a very careful thermal treatment for crystallization and melting. A much more diffuse fusion process is observed for a crystalline polymer which is polydisperse in molecular weight and which is heated very rapidly. The melting range now extends over many degrees and the melting temperature is not nearly as well defined as in Fig. 10. In the early studies of polymers, before their molecular constitution was recognized, the aforementioned kind of fusion curve was usually observed. Since it is almost a dogma in chemistry that pure substances must melt sharply it was concluded that these materials were not pure. It was further argued that since they were not pure substances they should not be studied. Consequently they were discarded. In retrospect it has become abundantly clear that these were indeed chemically pure species. The broad melting range can be properly attributed to morphological rather than chemical impurities. The small crystallite sizes, their distribution and the influence of the diffuse interfacial zone all contribute to the spreading out of the melting process. Superficially, the results are the same as if chemical impurities were present.

In contrast, the deliberate introduction of copolymeric units into a homopolymer chain influences melting in a very definite and systematic manner. We encounter here a phenomenon unique to long-chain molecules. The changes in the melting behavior depend not only on the amount and type of counits introduced but also on how they are distributed along the chain. For example, if the counits are randomly distributed then a systematic depression of the melting temperature is found and the fusion process becomes very broad. However, if the counits are introduced in long sequences, forming a block copolymer, then the melting temperature is invariant with compositional changes and the fusion process remains quite sharp. One then has a very potent tool with which to control chemical and physical properties of crystalline polymers. If, for example, chemically more reactive groups are desired

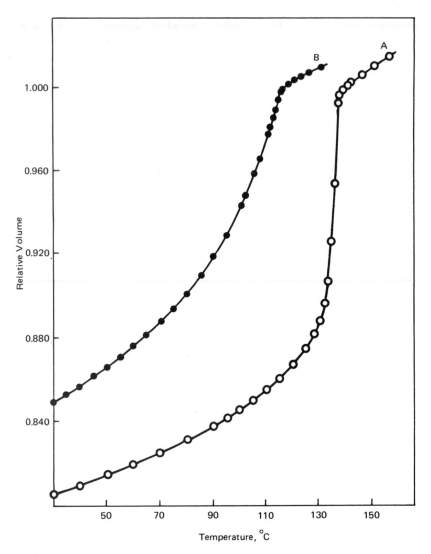

Fig. 11 Plot of relative volume against the temperature for linear polyethylene, curve A, and branched polyethylene curve B. From Journal of the American Society *75*, 4093 (1953). Reproduced with permission.

in a chain while it is necessary to maintain the melting temperature at the high value of the homopolymer the counits will be incorporated in an ordered manner. It may, however, be desirable to reduce the melting temperature for processing purposes. The random introduction of counits does this.

Other kinds of chain irregularities such as geometric or stereo-isomers, influence the melting temperature in a similar way. Thus, for example, while either highly isotactic or syndiotactic polypropylene forms a crystalline solid, when the stereo-isomers are randomly distributed along the chains a viscous liquid results. The presence of branched units also influences the melting behavior. Branching is characteristic of low density (as opposed to linear) polyethylene. This is a polymer that we have become accustomed to in our everyday life. A comparison of the melting of branched and linear polyethylene is given in Fig. 11. Although chemically identical, the melting point of the branched polymer is about 20 to 30 degrees lower than its linear counterpart. More striking, and more important from a practical point of view, is the much lower level of crystallinity characteristic of the branched polymer. Fusion takes place over a very wide temperature range, starting almost at room temperature, as compared to the very sharp melting of the linear polymer. These differences are naturally reflected in different physical properties and thus to the uses of these polymers.

Since the melting temperature of a homopolymer is a well-defined quantity and is easily determined it is of interest to see how it is influenced by the chemical nature of the repeating unit. Table I presents

Table I
Melting Temperatures of Homopolymers

		$T_m°C$
Polydimethyl siloxane	Silicone rubber	-40
Polyisoprene—1,4 *cis*	Natural rubber	28
Polyisoprene—1,4 *trans*	Gutta-percha	75
Polyisobutylene	Butyl rubber	5
Polyethylene—linear		145
Polyethylene—branched		115–120
Polytetrafluoroethylene	Teflon	330
Polypropylene		176
Polyacrylonitrile		317
Polymethylene oxide	Delrin	180
Polyhexamethylene adipamide	Nylon 6,6	270
Polyethylene terephthalate	Dacron	260
Cellulose triacetate	Acetate rayon	297
Cellulose trinitrate		>700

a summary of the melting temperatures of some representative polymers. These data clearly indicate that the melting temperatures vary widely with chemical type. In this respect we find no difference between polymeric and monomeric substances. There are also many examples in the table that show how the location of the melting temperature governs the end-use of the polymer. For example, polymers commonly used as elastomers must have low glass temperatures. If they possess a regular chain structure they must also have a low melting temperature. Otherwise the conformational requirements necessary for the elastic behavior will not be satisfied. We therefore find that three widely used elastomers—natural rubber, polyisobutylene, and polydimethyl siloxane—possess melting temperatures of 28°, 5°, and —40° respectively.[1] Thus, at room temperature they can be employed as rubbers. At the other extreme, polymers used as fibers must have very high melting temperatures. Two polymers extensively used as synthetic fibers, polyhexamethylene adipamide and polyethylene terepthalate, have very high melting temperatures as does polyacrylonitrile. The derivatives of cellulose are all very high melting. Cellulose trinitrate, used in gunpowder and in solid rocket propellants, has the highest known melting temperature of a polymer.

Polymers commonly used as plastics—such as polyethylene, polypropylene, and polymethylene oxide—have melting temperatures between those of fibers and elastomers. These polymers can be processed and fabricated in the molten state. When actually used they are in the crystalline state, so that the desired physical properties can be exploited. Such polymers are also called thermoplastics. The replacement of all the hydrogen atoms from linear polyethylene with fluorine atoms, to give polytetrafluoroethylene, increases the melting temperature by about 200°C. This change in chemical constitution yields one of the most thermally stable polymers known.

A detailed analysis of the molecular factors that determine the melting temperature has led to a causal relation with the disordered chain conformation. Generally speaking, the more flexible compact chains have the lower melting temperatures. The more rigid extended chains possess much higher melting temperatures.

Fibers

Fibers can be thought of as a special class of crystalline polymers. The crystalline regions in fibers are arranged so that the molecular chain axes are preferentially aligned, or oriented, in a direction parallel to the macroscopic fiber axis. This structural feature is common to all fibers whether they are synthetic or "man-made," as nylon and dacron, or of natural origin as wool and silk. The outstanding physical properties of a fiber are its very high tensile strength, i.e., the ability to sustain

---

[1] Although the melting temperature of natural rubber appears to contradict the principle just stated, it supercools very easily. Hence it is in the disordered, amorphous state at room temperature.

large forces without breaking, coupled with a low deformability. These properties are quite different from that of a rubber-like material and arise because of the major differences in chain structure and molecular organization.

A bulk synthetic polymer, usually in the form of a powder, is converted to a fiber by a technological process known as spinning. The first step in spinning requires either the melting or the dissolution of the polymer. Polymers that can be melted without molecular degradation are melt spun. Common fiber-forming polymers in this category are polyhexamethylene adipamide (nylon), polyethylene terephthalate (Dacron), and polypropylene. Because they are susceptible to thermal degradation, cellulose, cellulose acetate and polymers based on polyacrylonitrile are solution spun since the temperatures required are much lower than for melt spinning. The principles involved in these spinning processes are very similar. We therefore will focus attention only on melt spinning.

A schematic diagram of the apparatus involved in melt spinning is given in Fig. 12. Polymer, in the form of powder or chips, is fed from a hopper into the inert atmosphere of an electrically heated chamber maintained at a temperature above the melting temperature. The disordered polymer, in the liquid state, is then pumped through a large number of very small holes that are drilled into a flat plate. This plate, called a spinneret, is shown in the inset of Fig. 12. The liquid polymer emerges from the spinneret, in the form of thin threads, into air at a temperature well below the melting temperature. Crystallization naturally will ensue. At this stage in the process, however, the crystallite orientation is very poorly developed. The necessary high orientation for fiber properties is achieved by stretching, or "drawing," the crystalline threads. This is accomplished by passing the fibers through two pairs of rollers rotating at different speeds. The second roller is operated four to five times faster than the first one. This arrangement yields a net tensile force that stretches the fiber several hundred percent. During drawing, the temperature between the two sets of rollers is kept above the glass temperature but below the melting temperature of the polymer. The spinning process yields a very high crystallite orientation in a fibrous polymer. It does so in a very effective manner since a large quantity of fiber can be continuously produced.

The high state of crystallite orientation is evidenced by wide angle X-ray patterns which are similar to those illustrated in Fig. 4c. However, despite the great ingenuity that has been applied and the mechanical devices that have been developed for orienting crystalline polymers the extent and perfection of the orientation is not yet as great as is found in the naturally occurring fibers such as silk, wool, and cotton. The spinning processes and spinnerets of nature have not as yet been equaled

Fig. 12 Schematic diagram of melt spinning process.

or surpassed by man. In synthetic fibers the lamella-like crystallite structure must obviously have been reduced to some extent. The details of the morphological structure are still being debated. It is clear, however, that a fiber must consist of a periodic repetition of crystalline and amorphous zones. These structural features must of course turn up in the mehanical properties of the fibers.

A typical stress-strain curve for a synthetic fiber is given in **Fig.**

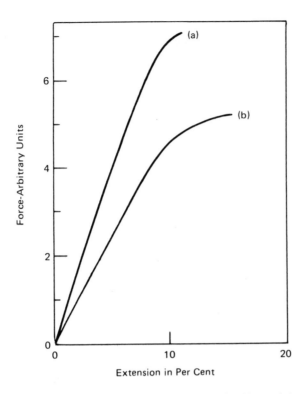

Fig. 13 Typical force-length curves for synthetic fibers. (a) polyethylene terephthalate (b) nylon.

13. These curves are quite different from that of rubber. The long-range elasticity is clearly lost, with the maximum extension being relatively small, on the order of 15 to 30 percent. The forces required for these extensions are, however, very large. The initial portions of the curves, representing the first few percent of extension is a result of the deformation of the noncrystalline chain units. Further deformation takes place at the expense of more orientation and eventually the destruction of the crystalline regions. This leads to a very steep stress-strain curve. The high stresses involved eventually lead to chain rupture and fracture of the sample. The important physical properties of fibers are high strength accompanied by a high modulus of elasticity, i.e., resistance to deformation. These properties are directly related to and are a consequence of the chain structure and morphology. The highly ordered arrangements of the chain atoms with the accompanying loss of conformational freedom yields a high modulus of elasticity and high strength.

These fundamental properties are not directly derived from any particular chemical structure. However, for a fiber to be useful as a textile material other conditions also have to be satisfied. For example, to be able to undergo ironing and washing a fiber forming polymer must have a high melting temperature so that it remains intact during these processes. Consequently, as we can tell from the data in Table I melting temperatures of at least 200°C are necessary to form a satisfactory fiber. The melting temperature is of course determined by the chemical nature of the repeating unit. To be attractive a fiber must be capable of accepting various kinds of dyes. This condition also limits the types of chemical repeating units that can be used. Copolymeric units are very often introduced to improve the dyeing capability.

It seems appropriate at this point to summarize some of the mechanical properties of polymers and to review how they depend on the different states. The hardness or deformability of a substance can be quantitatively represented by its Young's modulus, which is the ratio of the applied stress (force/unit area) to the relative extension (strain). The chart of Fig. 14 gives some typical values for polymers and monomers. The relative ease of deformation is inversely related to the Young's modulus. An enormous range in properties is represented in Fig. 14 even if

Youngs Modulus dynes/cm^2

Fig. 14 Youngs modulus for different substances.

considerations are restricted to polymeric substances. Crystalline, nonpolymeric materials, such as diamond and metals, represent the least deformable substances known. They are closely followed by inorganic glasses whose moduli lie in the range 10^{11} to 10^{12} dynes/cm.^{2} The moduli of polymeric fibers are in the lower portion of this range. The moduli progressively decrease for organic glasses, unoriented crystalline polymers followed by rubbers, which have the lowest value. The magnitude of the modulus must be intimately related to the molecular mechanisms responsible for the elastic response. This in turn is dependent on the molecular and morphological structures present. From our knowledge of polymer conformations and of the different crystallite arrangements that can exist the large variation in the modulus should not surprise us. The inelasticity of diamond and similar materials results from the fact that the constituent atoms are very firmly and rigidly bound to one another in a crystal lattice and can therefore not be easily displaced.

Table II
Tensile Strength of Materials

Materials	Tensile Strength
	$dynes/cm^2$
Steel, piano wire	2×10^{10}
Nylon, fiber	1×10^{10}
Cotton, cellulose	5×10^9
Viscose rayon, textile fiber	3×10^9
Wool	2×10^9
Nylon, crystalline unoriented	8×10^8
Polymethyl methacrylate, glass	8×10^8
Rubber	3×10^8

Another physical property of great practical importance is tensile strength, the ultimate load that a material can bear without breaking. Some typical values of this quantity are given in Table II. The strength of the materials order themselves in the manner that we would expect. The range in values, however, is not nearly as great as for the elastic modulus. The strongest materials are usually the least deformable. Therefore, one must always seek a compromise between high strength and high elasticity. The strength of a material is directly related to the molecular order and crystallite orientation. Crystalline, but unoriented nylon is only slightly stronger than rubber. However, when the crystallites are oriented, as in a fiber, the tensile strength increases more than 10-fold. Less oriented fibers such as cotton and viscose rayon have a somewhat lower tensile strength. The ultimate strength data, given in Table II, have been calculated on the basis of the cross-sectional area of the sample. On this scale the tensile strength of an oriented synthetic fiber is about half that of a high-strength steel wire. However, there is a large difference in the density (mass/unit volume) of the two materials. Steel is almost seven times denser than nylon. Therefore, when tensile properties are compared on a mass basis the nylon fiber is about three and one-half times stronger than the steel wire. This fact highlights one of the most important properties of polymeric materials. Namely, that when compared on a mass basis they are very strong indeed. The high strength of polymeric fibers can be illustrated by the fact that a 1-inch diameter nylon rope can easily lift a $5\frac{1}{2}$-ton load.

Mechanochemistry

In addition to the desirable mechanical and physical properties that make polymeric fibers so useful, they also possess another very important characteristic. This is a dimensional property that allows all fibrous macromolecules to be contractile, i.e., to undergo anisotropic dimensional changes. The fundamental reason for this behavior is the molecular dimensions that are associated with the distinctly different conformations of the crystalline state as compared to the liquid or disordered state. We recall that in the liquid state, where the chains are in statistical conformation, a linear dimension of a chain molecule is conveniently represented by the root-mean square distance between chain ends, $\langle r^2 \rangle^{1/2}$. This dimension is directly proportional to $n^{1/2}$, the square root of the number of bonds making up the chain. On the other hand, in the crystalline state, with an ordered chain conformation, a linear molecular dimension is proportional to n. For a high molecular weight chain, the ratio of these two dimensions can be quite large. For example, for a chain comprised of 10^4 bonds a 100-fold decrease in molecular length will occur upon disordering.

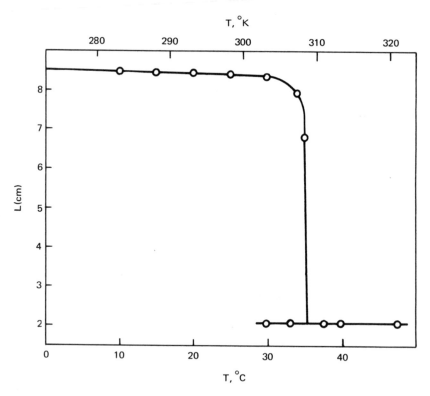

Fig. 15 Length-temperature relation for fibrous natural rubber under zero force. From J. F. M. Oth and P. J. Flory, Journal of the American Chemical Society *80*, 1297 (1958). Reproduced with permission.

The question then arises as to whether these molecular dimensional changes can be manifested in a gross macroscopic change during the transformation from one state to the other. Again a distinction must be made with respect to the nature of the crystalline state. If the crystalline regions are randomly arranged relative to one, another the local dimensional changes will average out. Therefore, the expected macroscopic changes will not occur. However, when the molecular axes of

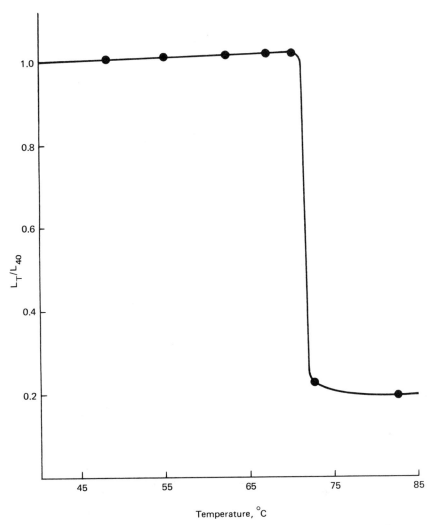

Fig. 16 Relative length-temperature relation for rat tail collagen under zero force. From P. J. Flory, Journal Cellular and Comparative Physiology *49*, (Supplement 1) 175 (1957). Reproduced with permission.

the ordered chains are preferentially oriented along a macroscopic dimension in space, as in a fiber, then the expected dimensional changes are indeed observed. The magnitude of this change will depend on the extent and perfection of the orientation. This dimensional property is what causes the shrinkage of wool and other synthetic fibers. In the laboratory this phenomenon can be studied more quantitatively. In Figs. 15 and 16 length-temperature relations are given for two very different polymers, fibrous natural rubber and the fibrous protein collagen. Although these fibers are very different chemically and their use is quite different, structurally they are both crystalline and axially oriented. Consequently, upon melting a large diminution in length is observed for both polymers. This change occurs very suddenly over a very small temperature interval. The plots in Figs. 15 and 16 have similar characteristics to the volume-temperature plot of Fig. 10. In the latter case the volume increases over a very narrow temperature interval. The razor-like sharpness of the change, whether it be in volume or in length, is a consequence of the fact that a melting transition is taking place.

The dimensional changes that accompany the melting process will

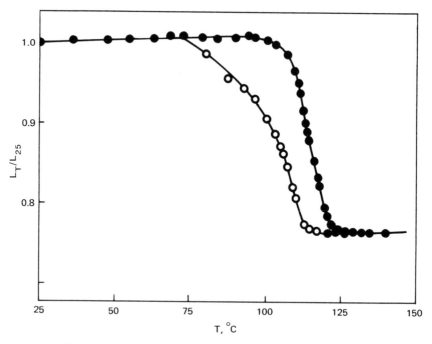

Fig. 17 Relative length-temperature relation for reversible contractile polyethylene fiber. ● heating; ○ cooling. From Journal of the American Chemical Society *81*, 4148 (1959). Reproduced with permission.

be reversible if an equilibrium force is imposed on the system or if an adequate number of intermolecular cross-links are introduced into the original fibrous structure. In Fig. 17 we have an example of a reversible, contractile polyethylene fiber. There is no external force acting on the fiber but it has been cross-linked. On the heating cycle shortening is observed concomitant with melting; on cooling and recrystallization re-elongation to the original dimensions takes place.

Fig. 18 illustrates the length-temperature relations that will be observed under a constant force. The melting temperature will increase

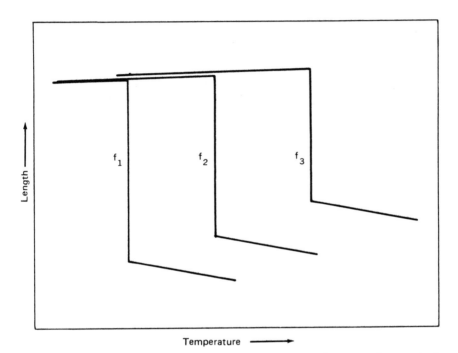

Fig. 18 Length-temperature relation for fibrous polymers under constant force f. $f_3 > f_2 > f_1$. Adapted from P. J. Flory, Science *124*, 53 (1956). Reproduced with permission.

with an increase in the applied force.[2] Shrinkage occurs over a very narrow temperature range but the magnitude of the change will decrease with an increase in the force. With this information we can now consider another problem: If the length is to be held constant, what force must be imposed, at a given temperature, to satisfy this condition? The answer

[2] This is completely analogous to the increase in the melting temperature that is observed for virtually all substances, with an increase in the applied hydrostatic pressure.

is that in order to prevent fusion and shrinkage the force must just exceed the equilibrium one. This force is known as the isometric force and is equal but oppositely directed to the force developed by the fiber upon melting. For synthetic fibers, isometric tensions as high as 4 kg/cm² have been observed. This tension is of the same order of magnitude as is developed by natural muscle systems in tetanic contraction.

In the examples illustrated melting and the accompanying dimensional changes, or tension development, were caused by changes in temperature. Identical processes can, however, be carried out at constant temperature. They require chemically more reactive fibers that can interact with appropriate reagents. Isothermal chemical melting, accompanied by corresponding changes in dimensions and in tension, has been amply demonstrated with the fibrous proteins. We give, as an example in Fig. 19, the shortening of glycerinated muscle fiber, a conventional muscle model, when immersed in a solution of adenosine triphosphate (ATP). In complete analogy with the previous figures a major diminution in

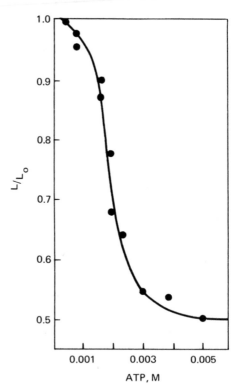

Fig. 19 Relative length change of glycerinated muscle fibers as ATP concentration is varied. From Proceedings of the National Academy of Sciences *45*, 814 (1959). Reproduced with permission.

length occurs with a small change in ATP concentration. This dimensional change is accompanied by the loss of the ordered fibrous structure. When experiments are conducted at constant length large tensions develop at the same ATP concentration that the major shortening occurs in Fig. 19. There is, therefore, a very simple and general process, applicable to all fibrous macromolecules, that allows for anisotropic dimensional changes and tension development on a macroscopic level. The only structural requirement is that the system be fibrous.

Contractility and tension development are very important phenomena that are observed in many natural occurring processes such as muscular contraction, cell motility and cell mitosis. These processes invariably involve the utilization of macromolecules. The question of whether the mechanism outlined is operative in these natural systems of great biological interest is the subject of intensive investigations at many laboratories.

Chapter VI

MACROMOLECULES OF BIOLOGICAL IMPORTANCE

Introduction

In this chapter we shall discuss the structure and properties of macro-molecules of biological and biochemical importance, primarily proteins and nucleic acids. Proteins can be thought of as naturally occurring copolymers of L-amino acids. In contrast to the usual synthetic copolymers, however, the sequential arrangement of the repeating units is very precisely and exactly defined. The particular arrangement leads in turn to a very definite structure that controls the function of the protein.

Proteins, which are of prime importance to life (the word "protein" comes from the Greek *proteios* meaning first rank), serve in many diverse roles. For example, they form the structural materials that hold living organisms together, they transport ionic and molecular species within a living system, they control and catalyze virtually all metabolic processes, and as antibodies they combat viral and bacterial diseases. A staggering number of different proteins is found in all forms of animal life.

Such great versatility in the behavior of a given type of macromole-cule may at first be surprising. However, 20 different amino acids, or counits, are found in proteins. Hence, even for a chain containing only 100 repeating units, it is possible to have 20^{100}, or 10^{300}, chemically distinguishable molecules. When the conformational versatility of the chain is coupled with these vast chemical possibilities, the great diversity of structure becomes apparent. In fact, only a very small fraction of the theoretical number of distinct molecules is actually found in nature. Before engaging in a more detailed discussion of protein structure and function, it is instructive to examine the properties of homopolymers formed from the L-amino acids. These polymers are called polypeptides.

Polypeptides

The repeating unit of a polypeptide chain is an amino acid whose skeletal structure is given in Fig. 1. In this monomer the α-carbon atom, C^α, is flanked by an amino group (NH_2) on one side and a carboxyl group (COOH) on the other. A hydrogen atom and an R group are

Fig. 1 Amino acid structure.

attached to the α-carbon. The R group represents a particular substituent or side chain. Twenty different kinds of R groups, and thus amino acid repeating units, are found in proteins. In solution, the amino acid can exist as a doubly charged species or zwitterion. There is a positive charge on one end of the molecule and a negative one on the other.

Because there are four different groups attached to the α-carbon (except when R = H), the amino acids can exist in two stereo-isomeric forms which are mirror images of one another. The two isomers are shown in Fig. 2. Of the two possible ways that the groups can be attached

L–Amino Acid   D–Amino Acid

Fig. 2 The two stereo-isomers of the amino acids.

to the α-carbon only one, the L isomer, is found in proteins. Although the other isomer is occasionally found in other naturally occurring molecules, it has not as yet been found in a protein. The reason for this strong preference for the L isomer is not as yet clear. The answer is clearly involved with the question of the origin of life itself.

Structural formulas for the 20 amino acids that constitute the repeating or counits of proteins are listed in Table I. Their formal names

Table I

Amino Acids Found in Proteins

Amino acid	Chemical Structure	Amino Acid	Chemical Structure
Glycine (Gly)		Threonine (Thr)	
Alanine (Ala)		Tyrosine (Tyr)	
Valine (Val)		Phenylalanine (Phe)	
Leucine (Leu)		Tryptophan (Try)	
Isoleucine (Ileu)		Aspartic (Asp)	
Serine (Ser)			

98

Table I (continued)

Amino Acid	Chemical Structure
Glutamic (Glu)	
Lysine (Lys)	
Arginine (Arg)	
Histidine (His)	
Methionine (Met)	

Amino Acid	Chemical Structure
Cysteine (CySH)	
Cystine (Cys)	
Proline (Pro)	
Hydroxyproline (Hypro)	

as well as the commonly accepted abbreviations are also given. The most striking feature of the table is the very wide range of chemical structures that are represented. These range from hydrocarbon type side chains (such as Gly, Ala, Val, and Leu) to those containing aromatic groups (such as Phe). Included are hydroxyl (OH)-containing residues (such as Ser and Thr) and charged acidic and basic groups (such as Asp and Glu and His, Lys and Arg). Proline and hydroxyproline are special, since the side groups are five-membered rings that include the α-carbon and adjacent nitrogen atoms. Cystine plays a unique role because the disulfide bond can act as a cross-linker between chains or between units of the same chain. In addition to the statistical feature of distributing the amino acid residues along a chain, the very distinct and different chemical properties are also important in determining structure and function. Since water is the natural habitat of protein molecules, we must recognize that the polar and charged groups show a strong desire to mix with water. These are termed hydrophilic groups. On the other hand, the hydrocarbon and nonpolar side groups would prefer to be removed from the aqueous medium. Such groups are called hydrophobic.

The formation of a polypeptide chain from the amino acid monomers is conceptually similar to the classical condensation polymerization. The monomer is bifunctional, and one group can react with the other while the bifunctionality remains intact in the resulting molecule. Hence, by the principles developed in Chapter II, long-chain molecules can be formed as is shown in Fig. 3. This reaction yields a dipeptide with an

Fig. 3 Formation of a dipeptide from two amino acids.

amino and a carboxyl end group. Further stepwise reactions yield long-chain polypeptides with the same terminal groups as is shown in Fig. 4. A large number of different copolymers can be devised depending on the chemical nature of the R groups and their sequential arrangement along the chain.

Fig. 4 Formation of polypeptide chain.

Conformational Properties of Polypeptides

From the skeletal chain structure in Fig. 4 we can see that there are three different bonds that span the distance from one α-carbon atom to the next one. The rotational characteristics of the C^α—C bond, the C—N bond, and the N—C^α bond are, therefore, very important in determining the conformational properties of the chain. The electronic structure of the C—N, or amide bond, is such that it possesses a great deal of double bond character. In effect, this bond is nonrotatable. This characteristic leads to a planar amide group with the sequence of atoms from one α-carbon to the next, the grouping C^α—CO—NH—C^α being constrained to lie in a plane of considerable rigidity, such as is shown in Fig. 5. However, rotations are still possible about the C^α—C and

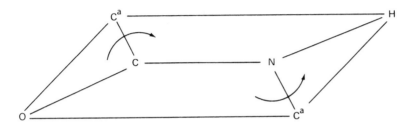

Fig. 5 Planar amide group.

the N—C^α bonds without altering the required planarity. The polypeptide chain will thus still possess a great deal of conformational versatility. The α-carbon atoms can be thought of as junction points that allow for the orientation of one planar group relative to its neighbor.

Polypeptides can thus adopt statistical conformations in the same manner as other chain molecules. The characteristics ratio, $C = \dfrac{\langle r^2 \rangle}{nl^2}$

will be expected to depend on the chemical nature of the side group. For polyglycine $(R = H)$, the characteristic ratio is 2. This low value indicates a very compact chain having considerable rotational freedom about the two single bonds. On the other hand, when bulkier side groups are present, such as in poly-L-alanine $(R = CH_3)$, the characteristic ratio increases to about 9, indicating a more extended chain structure. The poly-L-proline chain represents a special situation as can be seen from the skeletal structure given in Fig. 6. Here the R group is a five-membered ring, and one of the ring bonds is part of the main chain. The C—N bond still possesses its double bond character and is essentially nonrotatable. Since the N—C^α bond is part of the ring, its rotational

Fig. 6 Skeletal structure of poly-L-proline.

angle is also fixed. Hence, for this chain, rotation is only allowed about the C^α—CO bond. This rotation will, however, be severely restricted because of the steric interactions of the two adjacent rings. A highly extended molecule can be expected and indeed is found since the characteristic ratio for poly-L-proline is about 20.

When a preferred set of rotational angles are repeated successively along the chain, then an ordered conformation is generated, as we have found for other polymers. The ordered polypeptide structures can actually be observed in the isolated molecules in dilute solution. Polypeptides form three known major types of ordered structures.

The most widely known ordered structure is the α-helix of Pauling and Corey. Fig. 7a shows how a single polypeptide chain can be twisted in such a way as to form the helical structure of Fig. 7b while the planarity of the amide groups is maintained. Fig. 7b is a model of the α-helix, and the salient dimensional and geometric features of this conformation are shown in Fig. 8. The α-helix is right-handed, i.e., the chain backbone follows the pattern of the thread of a right-hand screw. An exact repeat of the structure occurs every 18 repeating units (or amino acid residues). This repeat corresponds to five turns of the helix or a linear advance of 27 Å. Each turn of the helix thus involves 3.6 residues, so that the helix is nonintegral. The recognition that nonintegral helices were permissible structures provided the necessary insight into solving this particular structural problem. This solution in turn opened the way for understanding the details of protein structure as well as the structure of other macromolecules pertinent to molecular biology.

A very important feature of the α-helix is indicated by the dashed lines of Fig. 7b. The spatial positions of the atoms are such that the hydrogen atom from a N—H group forms a hydrogen bond with the oxygen atom of the C=O group on the third amino acid residue down the chain. In this structure the bonding atoms find themselves the proper distance apart on successive turns of the helix. This *intramolecular* hydrogen bond is a feature unique to the α-helical structure, giving it a very

Fig. 7 Schematic representation of twisting an extended polypeptide chain into the α-helical conformation.

stable conformation. A well-defined, ordered structure can thus be formed and stabilized within a single polypeptide chain without any intermolecular interactions being involved.

The fact that polypeptide chains can form an ordered helical structure should not surprise us. We have already encountered this kind of structure in our study of synthetic polymers, particularly those that are stereo-regular and isotactic. From purely a geometric point of view, without concern for specific atomic positions and interactions, there is a great deal of structural similarity between the two types of chains as is seen in Fig. 9, which compares the α-helix formed by polypeptides and the 3.5 helix (3.5 monomer units per turn) generated by an isotactic polymer. The helical forms are very similar despite the great difference in the atomic constitution of the two kinds of molecules.

Another important ordered structure generated by polypeptides is one in which the chains are in a much more extended conformation. This is called the β-conformation, and the chain is almost in planar

Fig. 8 Detailed geometry of α-helical conformation. Reproduced with permission from C. B. Anfinsen *The Molecular Basis of Evolution,* John Wiley and Son.

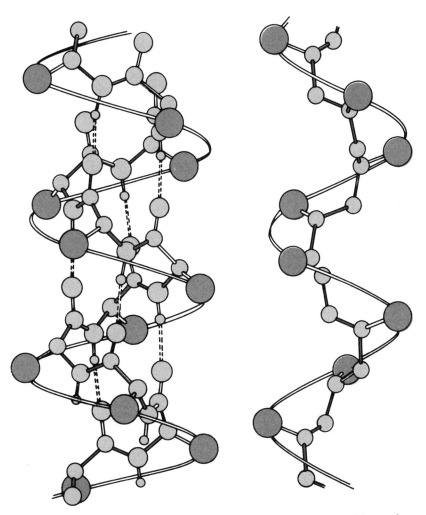

Fig. 9 Comparison of the α-helix and helical structure generated by an isotactic vinyl chain. From G. Natta and P. Corradini, Rubber Chemistry and Technology *33*, 703 (1960). Reproduced with permission.

zigzag form. The structure of a single chain in this conformation is shown in the left-hand portion of Fig. 10. Such chains are very often organized in sheet-like structures, as is indicated on the right side of this diagram. Alternate chains can be placed either parallel to one another or can be oppositely directed or antiparallel. The distance between the chains is such that in order to achieve the most effective intermolecular hydrogen bonds (indicated by the dashed line in the figure), the

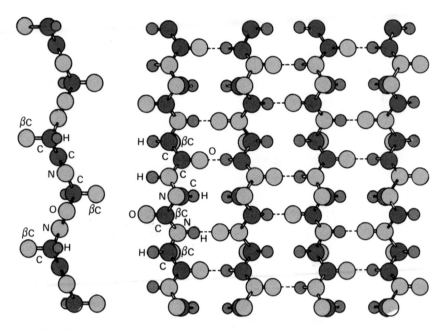

Fig. 10 The β conformation of polypeptide chains. From R. E. Marsh, R. B. Corey and L. Pauling, Biochemica et Biophysica Acta *16,* 13 (1955). Reproduced with permission.

individual chains are slightly pleated rather than being fully extended. The β-structure of the polypeptides is very similar to the ordered structure of the synthetic polyamides.

The third ordered structure is unique to poly-L-proline and results from the special chemical constitution of this polymer. This is also a helical structure, but it has a left-handed sense as compared with the right-handedness of the α-helix. A schematic diagram of this structure is given in Fig. 11. This is an integral helix with three residues per turn. In contrast to the α-helix, the C=O groups are now directed almost perpendicular to the helical axis. Because of this orientation, these groups cannot interact with others along the chain. The stability of this helical structure arises solely from steric constraints.

In order to generate ordered polypeptide structures, it is not mandatory that the chain have chemically identical repeating units. Since the chain backbone is the same for all the amino acids, units with different side groups can be accommodated within the same structure. There are, however, some exceptions to this coordering or cocrystallization. For example, side groups having the same charge, and in close proximity, will repel each other and destroy the ordered structure. For steric reasons proline can only occupy the initial position of a right-handed α-helix

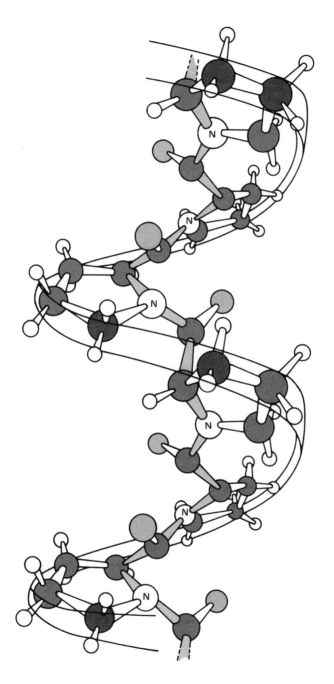

Fig. 11 Representation of the ordered conformation of poly-L-proline.

and cannot occur within the body of such a helix. Cystine, with its disulfide cross-link, is also very difficult to accommodate in any of the ordered structures. A similar restriction holds for other residues with very bulky side groups. In copolypeptide chains, ordered structural sequences can be interspersed with disordered ones, depending on the composition and distribution of the amino acid residues.

Proteins—General

A protein is a polypeptide chain, or collection of chains, that has a very definite sequence of amino acid residues. This specificity of repeating units, with different side groups, gives the molecule its structural form, which in turn determines its function. Since 20 different amino acids are available as counits, an astronomical number of chemically distinct molecules can be generated with a long-chain. Therefore, the same fundamental type of chemical repeating unit can participate in molecules that vary widely in their size, shape, and mass and also display a diversity of function. The range, scope, and vitality of proteins in man and lower organisms are truly remarkable. Proteins provide us with protection in the form of skin, hair, nails, and claws. Our connective tissue and tendons are proteins, as is the contractile apparatus of muscle and our veins and arteries.

Besides providing us with protection, physical structure, and mechanical capability, proteins are involved in a large number of chemical processes that are vital to our existence and well-being. In the form of enzymes they serve as catalysts for the biochemical reactions necessary to sustain life. They serve as carriers of oxygen and other small molecules and ions to various parts of the body. The globulins and antibodies with which we fight disease, are also proteins. Cell membranes and our mitotic apparatus are also made up of this class of macromolecules. We might then properly ask how this all comes about. The answer to this question lies in the fundamental principles of macromolecular properties which has been our major theme.

A given protein can be characterized by its overall composition, i.e., the relative proportion of the different amino acid residues that are present. More important, however, is the sequential arrangement of these units along the chain. Starting from one end of the chain, conventionally the amino end, (NH_2), one can specify in principle, and also in practice in many cases, each successive amino acid residue until the terminal carboxyl, (COOH) group, is reached. The overall composition and sequential arrangement are, of course, different for the different proteins. The rotational characteristics of the single bonds in the chain backbone, moderated by the spatial expanse of the side groups and their interac-

Table II
Amino Acid Composition of Selected Proteins*

Amino Acids	Human Insulin	Horse Hemoglobin	Egg Albumin	Ribonuclease	Silk Fibroin	Wool Keratin	Collagen
Molecular Weight	6,000	64,500	45,000	14,000	—	—	—
Glycine	4	48	19	3	581	87	363
Alanine	1	54	35	12	334	46	107
Valine	4	50	28	9	31	40	29
Leucine	6	75	32	2	7	86	28
Isoleucine	2	0	25	3	8	0	15
Phenylalanine	3	30	21	3	20	22	15
Tryptophane	0	5	3	0	0	9	0
Proline	1	22	14	4	6	83	131
Hydroxyproline	0	0	0	0	0	0	107
Serine	3	35	36	15	154	95	32
Threonine	3	24	16	10	13	54	19
Tyrosine	4	11	9	6	71	26	5
Aspartic Acid and Asparagine	3	51	32	15	21	54	47
Glutamic Acid and Glutamine	7	38	52	12	15	96	77
Lysine	1	38	20	10	5	19	31
Arginine	1	14	15	4	6	60	49
Histidine	2	36	7	4	2	7	5
Cysteine	0	4	5	0	0	0	0
Cystine	3	0	1	8	0	49	0
Methionine	0	4	16	4	0	7	5

* When the molecular weight is known, the composition is expressed as the number of amino acid residues per molecule. When the molecular weight is not known, the composition is expressed as the number of moles of amino acid residues per 100,000 g of protein.

tions, yield a unique, very specific three-dimensional structure. We shall soon discuss in detail several examples of three-dimensional protein structures. At present, it suffices to say that this well-defined structure establishes the function of the protein through shape specificity and the location of reactive side groups.

To start a more detailed examination of these principles, we list the amino acid compositions of some typical proteins in Table II. Virtually all of the amino acid residues are represented in each of the proteins. The overall composition varies widely, however, from one protein to the other. In some proteins all the amino acids are more or less equally well represented. In others, however, there is a very definite predominance of certain kinds of residues. We shall have occasion to refer to Table II, as well as to the compositional characteristics of other proteins, as we examine specific cases in more detail.

The exact sequence of amino acid residues has been determined for a large number of proteins. The task still remains to be accomplished for many more, particularly those of very high molecular weight. An example of the sequence analysis for the enzyme ribonuclease is given in Fig. 12. This protein, which has a molecular weight of 14,000, is

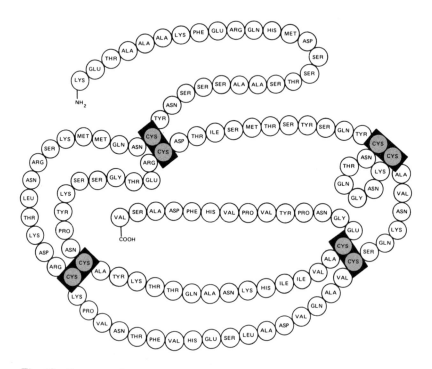

Fig. 12 Sequence of amino acids in ribonuclease.

composed of 124 residues arranged in the manner indicated. There are eight cystine groups in the single chain comprising the molecule. They are so positioned and the chain so organized that four pairs of disulfide (S—S) cross-links are made. In this case, the cross-links are intramolecular, i.e., within one chain or molecule. The diagram in Fig. 12 is not intended to represent the real structure. It is just a convenient device by which to convey the sequential order of the repeating units. In a very notable achievement, ribonuclease has been completely synthesized in the laboratory, starting with the monomeric amino acids. The 124 amino acids were assembled in the required sequence. The resulting synthetic polymer possesses all the structural, chemical, and enzymatic properties of the native protein. The path now appears to be open for the complete synthesis of other proteins.

Fibrous Proteins

Proteins can be conveniently divided into two categories: globular and fibrous. The fibrous proteins follow the same structural and morphological definition that we have already given to the synthetic fibers. They are crystalline, and the chains are preferentially oriented in an axial direction. The ordered conformations found in the fibrous proteins correspond to the major ones we have discussed for the polypeptide chains. This does not mean that the complete protein molecule is crystalline and ordered. Among the more important fibrous proteins are the collagens, which are the proteins of connective tissue, of tendons, and of hide. The keratins, which give protective covering in the form of skin, fur, hair, feathers, nails, and scales are also fibrous proteins. Myosin is a fibrous protein directly involved in muscular contraction. Other fibrous proteins are epidermin, another skin component, and fibrin, the protein that forms blood clots. We shall focus our attention on three of the major fibrous protein systems: wool or hair, silk, and collagen.

The α keratins are a very familiar class of fibrous proteins. Two typical members are wool and hair. An examination of the amino acid composition of wool keratin, from Table II, indicates the presence of most of the amino acids. There is, however, a very high concentration of cystine residues which is characteristic of the α-keratins and is indicative of a high level of intermolecular cross-linking. The primary structural unit of wool or hair is the α-helix. X-ray diffraction and electron microscope analysis make abudantly clear, however, that the ordered chains are not simply arranged side by side. Instead a very detailed organizational hierarchy exists.

The smallest organized entities that have been established are fibrous structures which lie along the complete length of the sample. These are about 20 Å in diameter and are called protofibrils. Each protofibril is probably composed of three α-helical chains twisted around each other so as to form a coiled-coil or supercoil. The chain structure within a protofibril is illustrated in Fig. 13. Each strand in this figure represents

Fig. 13 Schematic representation of three α-helical chains entwined around one another to form a protofibril of α-keratin. Adapted from L. Pauling and R. B. Corey, Nature *171,* 59 (1953).

an α-helix. The detailed organization of the protofibrils has been revealed by the electron microscope. As the cross-sectional view of Fig. 14 shows, nine protofibrils are arranged in a circle about two others. Thus, an 11-stranded cable, approximately 80 Å in diameter is formed. This structure, or microfibril, is embedded in an amorphous, nonorganized protein matrix. Hundreds of such microfibrils are then joined together to form a fibrous bundle, or macrofibril, which is about 2,000 Å in diameter. Thus, in the α-keratins there is an ordered progression of increasingly complex structures, starting with the α-helix of a single protein chain.

The physical properties of wool, hair, and the other α-keratins are derived from this morphological structure and from the peculiarities of the chemical composition. Within the protofibrils, the α-helices are intermolecularly cross-linked by disulfide bonds from the cystine residues. This represents a special feature of the α-keratins. The elastic properties of the keratins are qualitatively similar to the synthetic fibers and reflect the highly ordered chain structure. The extensibility is, however, tempered somewhat by the disulfide content. The low sulfur-containing keratins are more extensible than those containing large amounts. The latter are very inextensible.

The shrinkage of wool and hair when immersed in certain solvents is well known. Irreversible or irrecoverable shrinkage results from the permanent breaking of the intermolecular disulfide cross-linkages by these reagents. Reversible dimensional changes accompany the melting of the ordered structure and its recrystallization, while the intermolecular disulfide cross-links are maintained. This process follows the principles discussed in Chapter V.

The permanent waving of hair is based on the dimensional and elastic properties of the cross-linked chains. In the first stage of the process the disulfide bonds are severed. The enhanced elasticity allows the hair to be shaped or curled into the desired form. The disulfide bonds are then reformed giving a "permanent" shape to the hair. The original

Fig. 14 Electron micrograph of cross-section of an α-keratin fiber. Insert is a schematic representation of the electron micrograph results. Reproduced with permission from B. K. Filshie and G. E. Rogers, J. Molecular Biology *3*, 784 (1961).

process, developed many years ago, used curling irons whose high temperature caused the breaking of the cross-links. More sophisticated chemical methods have now been developed which are the basis for home permanents carried out at room temperature. One reagent severs the cross-links by chemical reaction with the disulfide bonds. The S—S bond is converted to two —SH species, separating the chains and allowing the hair to be more easily deformed. After the hair is set in the desired manner, this reagent is washed out. The addition of another reactant reforms some of the disulfide cross-links and gives a permanency to the structure.

In contrast to wool and the other α-keratins, there are no sulfur-containing amino acid residues in silk, which is composed mainly of Gly,

Ala, and Ser residues with relatively small amounts of the more bulky side chains. All silks are not, however, chemically identical. The different species of the silkworm produce a fibrous protein that contains different proportions of amino acids with bulky side groups. These compositional differences in turn influence the physical properties of the fibers. In the most common form of silk, from the silkworm *Bombyx mori*, there is a predominant six residue sequence, Gly-Ser-Gly-Ala-Gly-Ala, which repeats itself for long distances along the chain. This sequence accounts for a large proportion of the amino acid residues that are present.

The ordered portion of the polypeptide chains in silk are in the extended β-conformation illustrated in Fig. 10. Neighboring chains are antiparallel to one another, and a pleated sheet is formed. The Gly residues lie on one side of the sheet and the Ser and Ala on the other. Although the ordered structure of silk is quite well defined, it is not the only determinant of properties. The amino acids with bulky side groups cannot be accommodated within this three-dimensional, ordered structure. Therefore, disordered amorphous regions coexist with the crystalline ones. The constitution of silk is analogous to that of a block copolymer. Different classes of amino acids reside in each block. One type can participate in the ordered crystalline array while the other cannot.

The elastic properties of silk correlate very nicely with the fraction of bulky side groups present and thus with the crystalline-amorphous ratio. Force-length curves for three different types of silk fibers are shown in Fig. 15. Fibers from *Bombyx mori*, in which 13 percent of the residues have bulky side groups, are about 60 percent crystalline and can be extended 24 percent at its breaking point. On the other hand, in *Anaphe moloneyi*, only 5 percent of the residues are those with bulky side groups. It is , therefore, a highly crystalline, very rigid fiber that can be extended only 12.5 percent at its breaking point. The residues of *Antherea mylitta* contain 29 percent bulky side chains. It is the least crystalline of the silks and consequently can sustain the largest deformation. The same principles thus govern the physical properties of both the natural and synthetic fibers.

Collagen is one of the most abundant proteins in nature. In man it accounts for about one-third of the body protein. It is is the principal fibrous component of cartilage, ligaments, and tendons and is also found in hide and in the eye cornea. It serves as a matrix upon which the mineral hydroxyapatite crystallizes to form bone. The outstanding property of collagen is its rigidity and resistance to deformation. Thus, nature has put it to use in bone, in strengthening animal skin and hide, and in connective ligaments and tendons where the mechanical force generated by muscle must be transmitted without loss. When collagen is dissolved

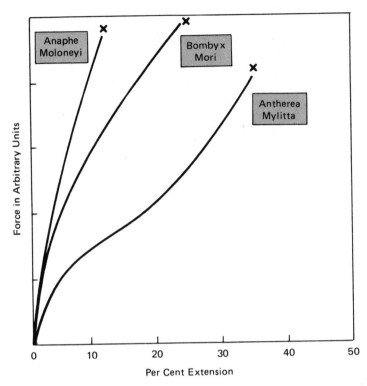

Fig. 15 Force-length curves for different silk fibers. Adapted from F. Lucas, J. T. B. Shaw and S. G. Smith, Shirley Institute Memoirs *28, 77* (1955).

in hot water, it is transformed into gelatin, which is much more soluble and is easily digestible.

We might expect that because of their special functions the collagens will have a unique amino acid composition. This turns out to be the case, as is revealed by the data in Table II. On a gross composition basis all collagens contain about 33 percent glycine. Furthermore, every third residue is glycine, so that that amino acid is uniformly distributed. Another distinguishing compositional feature of this protein is the very high content of proline and hydroxyproline. The concentration of these residues varies with collagen type and species and can be as high as 25 to 30 percent is some samples. Such high concentrations are unique among the proteins. Certain sequences of residues—such as Gly-X-Pro, Gly-X-Hypro, and Gly-Pro-Hypro—predominate in collagen. They form the basis for the particular ordered structure characteristic of collagen.

The fundamental structural unit of collagen is a very elongated molecule called tropocollagen. An electron micrograph of such molecules

Fig. 16 Electron micrograph of calf skin tropocollagen. Reproduced with permission from R. V. Rice, Proceedings National Academy of Science (U.S.) *46*, 1186 (1960).

is shown in Fig. 16. The tropocollagen molecule consists of three poly-peptide chains intertwined with one another in such a way as to form the super helix shown in Fig. 17. Each chain is essentially a polyproline left-handed helix (see Fig. 11) with three residues per repeat. However, they develop a gentle twist to the right when they interact and combine to form the triple-stranded super helix of tropocollagen. The spatial arrangement of atoms in this particular array is such that two out of every three amide (NH) groups and carboxyl (CO) groups hydrogen bond to one of the other two chains as is indicated by the dashed lines in Fig. 17. The distance between the chains and their periodicities are such that residues with side chains cannot be accommodated in every third repeating unit. This position must therefore be reserved for the Gly residue. Thus, the structure, composition, and distribution are consis-tent with one another. The bulky rings from the Pro and Hypro residues can fit into the structure without undue strain and without disrupting the hydrogen bonding pattern between chains. This particular ordered structure requires sequences of the type Gly-X-Pro, Gly-X-Hypro, and Gly-Pro-Hypro. Other sequences cannot be accommodated and would therefore not be found in ordered regions. The particular ordered struc-

Hydroxyproline

Proline

Glycine

Fig. 17 Schematic representation of the structure of tropocollagen. Three polypeptide chains are intertwined to give this structure. Reproduced with permission from J. R. Holum, *Principles of Physical, Organic and Biological Chemistry,* John Wiley and Sons, Inc.

ture of tropocollagen results from the unique amino acid composition and sequential distribution. In the molecular structure of tropocollagen one can already discern a source of the strength and rigidity of the collagen system.

The tropocollagen molecules form large macroscopic aggregates in different biological tissues. One kind of aggregate, found in skin and hide, is fibrillar as is shown in the electron micrograph of Fig. 18. These

Fig. 18 Electron micrograph of collagen fibrils. Reproduced with permission from R. V. Rice, Proceedings National Academy of Sicences (U.S.) *46,* 1186 (1960).

fibrils are about 5,000 Å in diameter and have a characteristic banded pattern which repeats every 640 Å. The banding results from the periodic absorption of the stain used in electron microscopy. This particular aggregate is formed by an overlapping set of perfectly aligned tropocollagen molecules, which are displaced one quarter of their length relative to one another.

The shrinkage of collagen fibers is a well known phenomenon that we have already discussed in Chapter V. A very strong correlation has been found between the shrinkage temperatures, in water, for collagens from many different sources and the proline and hydroxyproline content. The greater the concentration of these residues, the higher the shrinkage temperature. Upon further inquiry it is found that collagen molecules suitable to either cold or warm habitats are actually devised by nature through control of the amino acid composition. This concept becomes particularly striking when a comparison is made between cold-water and warm-water fish. Through control of the proline-hydroxyproline content, the collagen of warm-water fish have higher shrinkage temperatures than those whose natural habitat is the cold. The shrinkage temperature appears to be about 15° to 20° greater than the highest temperature likely to be encountered by a species.

Globular Proteins

It is evident from our discussion of the structure and properties of the fibrous proteins that they have very strong similarities with fibrous synthetic polymers. The globular proteins, however, present a unique molecular structure which we have not as yet encountered in our discussion of macromolecular systems. One of their distinguishing features is their compactness—virtually all the space within the domain of these globular molecules is filled and there is a negligible void volume. This structure is quite different from that of a chain molecule in a statistical conformation. The chain bonds do not adhere to a simple ordered conformation since the rotational angles do not repeat in a definite pattern. The structure of a given globular protein, although irregular, is also unique. Each globular protein represents a well-defined structure designed to carry out its specific function and mission. Polar, hydrophilic residues are found mainly on the exterior of the molecule, while the interior consists mostly of the nonpolar hydrophobic residues. The distribution of the residues is consistent with the fact that the natural habitat of the globular proteins is an aqueous medium.

The size, shape, and mass of the globular proteins vary over very wide extremes. Their molecular weights can be determined by standard methods. In some cases the molecules are so large that their outlines can be seen with the help of the electron microscope. The size and shape of the smaller molecules are deduced by methods of protein physical chemistry. Fig. 19 gives a qualitative description of the shape and dimensions of some typical globular proteins, as well as their molecular weights. Many of the globular proteins can be crystallized from solution. Here the individual molecules are organized to form a three-dimensional, ordered array. When the molecules can be crystallized, then the methods of X-ray diffraction can be brought to bear on the problem to give a very precise (within 1 to 2 Å) location of all the atoms. Thus, a very detailed structure of a very complicated molecule can be obtained.

In pursuing the premise that the sequence of amino acid residues determines the three-dimensional structure of a protein, we shall examine

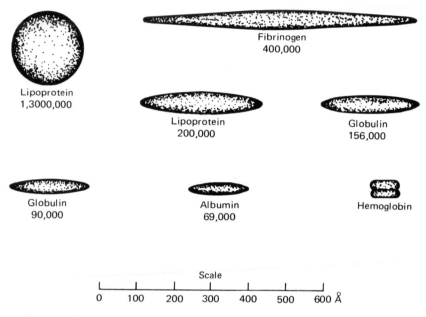

Fibrinogen
400,000

Lipoprotein
1,3000,000

Lipoprotein
200,000

Globulin
156,000

Globulin
90,000

Albumin
69,000

Hemoglobin

Scale

0 100 200 300 400 500 600 Å

Fig. 19 Shape, dimensions and molecular weight of several globular proteins. Reproduced with permission from E. R. Hardwick and C. M. Knobler, *Chemistry: Man and Matter,* Ginn and Company (1970).

this relation for some of the globular proteins whose structure has been elucidated by X-ray diffraction. The structures of several enzymes have now been detailed by this method. Enzymes are catalysts for biological and biochemical reactions. They behave as all other catalysts in that they increase the rate of a chemical reaction but are not consumed in the reaction. They are, therefore, very important in controlling chemical processes in nature.

One enzyme whose detailed structure has been elucidated is lysozyme, discovered by Alexander Fleming, who is also known for his discovery of penicillin. Lysozyme attacks many bacteria by dissolving, or lysing, their cell walls. This is accomplished by degrading a complex polymeric sugar or mucopolysaccharide. The enzyme catalyzes the breaking of specific chain bonds in the polysaccharide. Each molecule of lysozyme consists of a single polypeptide chain of 129 amino acid residues having a molecular weight of 14,600. There are four disulfide intermolecular cross-links in each chain. The sequential arrangement of the amino acid residues in the chain is shown in Fig. 20. One, of course, cannot deduce very much about the real structure from this planar display.

A space-filling molecular model representing the three-dimensional structure is shown in Fig. 21. The compactness of this molecule is readily

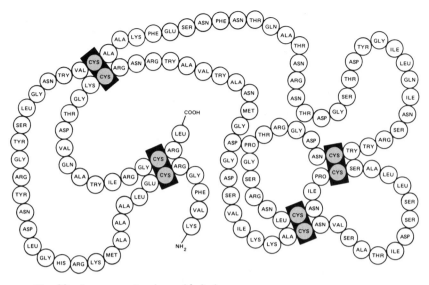

Fig. 20 Sequence of amino acids in lysozyme.

apparent. Most of the polar, hydrophylic side chains are found on the surface of the molecule and can thus interact favorably with the surrounding water. The hydrophobic, nonpolar side chains are in the interior, for the most part, and are thus shielded from the solvent. The two major ordered polypeptide conformations, the α-helix and the β-sheets are found in some portions of the molecule. Other portions of the chain do not, however, have any recognizable ordered sequence, and the molecule as a whole has its own very special structure. The consequence of the sequential arrangement is the structure represented in Fig. 21. The molecule is egg-shaped, and its most striking feature is the deep crevice running across its center. The whole strategy of stringing together this particular sequence of amino acids was to form this crevice, for it is here that the enzymatic action takes place. The crevice, or cleft, on the enzyme surface is so designed that a portion of the sugar molecule (the substrate) fits very snugly into it. There is a very good geometric fit between the reactant and a portion of the enzyme surface. This fit is demonstrated in Fig. 22 where the substrate has been placed into the catalytic trough. The amino acid residues that participate in the reaction are found at the catalytic site. In this particular case residues 35 (glutamic acid) and 52 (aspartic acid) are involved. The design of this molecule, through the particular repeat of amino acids along the chain, has been to have it soluble in an aqueous medium, to present a proper geometric surface or catalytic trough for the particu-

121

Fig. 21 Molecular model of the three-dimensional structure of lysozyme. Reproduced with permission from R. E. Dickerson and I. Geis, *The Structure and Action of Proteins,* Harper and Row, Inc. Illustrations copyrighted by R. E. Dickerson and I. Geis.

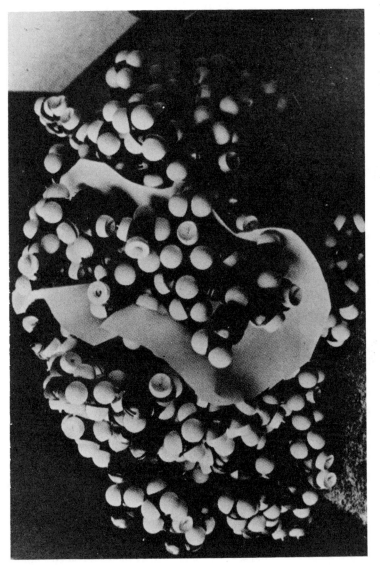

Fig. 22 Three-dimensional model showing fit of the substrate (sugar) into the catalytic crevice of lysozyme. From *Concepts in Biochemistry* by F. J. Reithel. Copyright 1967. Used with permission of Mc-Graw-Hill Book Company.

lar sugar, and to have accessible the appropriate functional groups to carry out the reaction. It is quite obvious that nature as a producer of polymers, designed for a specific chemical function, is far superior to the laboratory.

Ribonuclease is another enzyme that speeds up a chain-cutting reaction. However, instead of severing a polysaccharide chain, it performs the same function on a ribonucleic acid molecule (see next section). The amino acid sequence in ribonuclease has been given in Fig. 12. The three-dimensional arrangement of these sequences is shown in Fig. 23. In a general way the structure of this enzyme is similar to lysozyme. It is ovoid shaped with a deep crevice across its center. The crevice provides the proper geometrical fit for the nucleic acid. The functional groups necessary to carry out the scission are located at the surface of the crevice. These are residue numbers 12, 41, and 119. Although these residues are far apart in a linear array, they are brought relatively close to one another in the three-dimensional structure.

The enzymes whose detailed structures are known are all chain-cutting enzymes, and they have certain major structural features in common. They are compact, oval-shaped molecules with deep crevices, or depressions, so precisely designed that the reactant fits in and can react with the residues of the "active site." Many enzymes catalyze reactions that involve small molecules rather than polymer chains. The detailed spatial structures of these enzymes have not as yet been elucidated. However, one can reasonably expect that steric specificity will be the basis for the catalytic action here also.

In another set of functions the globular proteins act as transporters and storehouses for low molecular weight species. In vertebrates, one of the most important pairs of such molecules are hemoglobin and myoglobin. Hemoglobin, the main component of the red blood cells, binds molecular oxygen at the lungs and delivers it to myoglobin at the muscles. Here, the oxygen is stored until required for metabolic oxidative processes. On the return trip, the hemoglobin molecules help transport carbon dioxide back to the lungs. A single red blood cell contains about 280 million hemoglobin molecules each having a molecular weight of 64,500. The oxygen-binding capability of this molecule is due to the heme group associated with it. This group, illustrated in Fig. 24, consists of an iron atom surrounded by a porphyrin ring. The large number of double bonds in this ring results in a nearly planar heme molecule. The heme group also has characteristic colors. It is the oxygenated hemoglobin that is responsible for the deep red color of arterial blood; the bluish color of venous blood is due to the deoxygenated form of hemoglobin. In tissues, such as muscle, the red color is due to the oxygen attached to the myoglobin molecules.

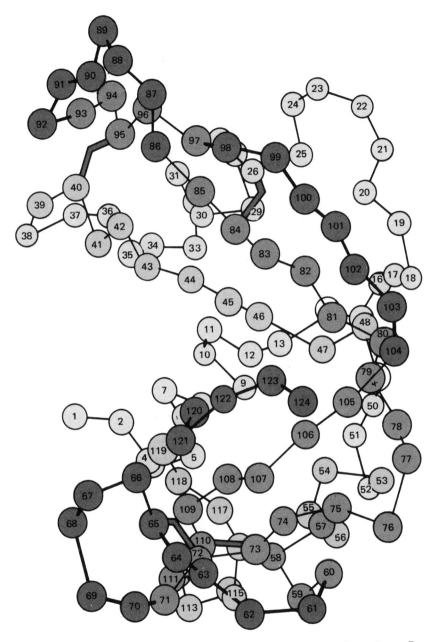

Fig. 23 Representation of three-dimensional structure of ribonuclease. Reproduced with permission from R. E. Dickerson and I. Geis, *The Structure and Action of Proteins,* Harper and Row, Inc. Illustrations copyrighted by R. E. Dickerson and I. Geis.

Fig. 24 Skeletal structure of heme group.

Myoglobin consists of a single polypeptide chain of 153 amino acid residues having a molecular weight of 17,000. A single heme group is tightly bound to the chain, and the molecule can reversibly bind oxygen. The detailed structure of myoglobin has been elucidated by Kendrew and is shown in schematic form in Fig. 25. The heme group is represented by the flat disc. The major function of the polypeptide chain is merely to serve as a container for this group. This protein has a very high proportion—about 80 percent—of its residues in the α-helical conformation. The container can be looked upon as being built up from eight connected pieces of α-helix.

Each hemoglobin molecule, the oxygen carrier, consists of two pairs of polypeptide chains. The α-chains each contain 141 amino acid residues, while the β-chains contain 146 residues each. The overall molecular weight of hemoglobin is 64,500. The four chains are so structured

Fig. 25 Schematic representation of three-dimensional structure of myoglobin. Reproduced with permission from *The Molecular Basis of Life,* W. H. Freeman and Co. Appeared originally in Scientific American.

that they each entwine a heme group. The detailed structural analysis of hemoglobin has been worked out by Perutz and coworkers. Because of their pioneering work in determining detailed protein structures by the methods of X-ray crystallography, Perutz and Kendrew received the Nobel Prize for Chemistry in 1962.

The hemoglobin structure is shown in Fig. 26. The oxygen-binding site is marked by O_2. Each chain has an irregular but well-defined structure very similar to that of myoglobin, and the chain pairs are symmetrically related. The packing of the four chains is such that there is close contact of side groups between unlike chains but virtually no contact between like chains. The overall molecular shape is roughly spherical. The hemoglobin molecule is more than just a receptacle for oxygen. It is so constructed that it can also control the amount of oxygen bound. All the details of the control mechanism have not as yet been worked out. However, it appears to involve a displacement of the β-chains. This movement alters the distance between the iron atoms, which in turn influences the amount of oxygen bound.

β_2 β_1

α_2 α_1

Fig. 26 Schematic representation of three-dimensional structure of hemoglobin. Reproduced with permission from R. E. Dickerson and I. Geis, *The Structure and Action of Proteins,* Harper and Row, Inc. Illustrations copyrighted by R. E. Dickerson and I. Geis.

Outstanding advances have been made in our understanding of the molecular basis of some diseases. What is involved here is the manufacture or synthesis of molecules that differ from the normal ones found in a healthy person. The hemoglobin molecules shown some very striking examples of molecular diseases. In some anemic patients there are abnormal α-chains or abnormal β-chains. The iron atom, which is usually in the ferrous state, is converted to the ferric state. When this occurs, the heme group loses its power to combine with molecular oxygen. This type of anemia is called ferrihemoglobinemia. Much more subtle differences in chain structure are also known to cause disease. For example, the sixth residue from the amino end in the β-chain of a normal hemoglobin is glutamic acid (Glu). Persons suffering from sickle cell anemia have valine (Val) substituted for Glu at this position. Glutamic acid has a carboxyl group, —COOH, on its side chain and thus can be ionized.

On the other hand, valine has a hydrocarbon isopropyl side group, which cannot ionize. When the same glutamic acid is replaced by lysine, thalassemia results. When glutamic acid in the 26th position is replaced by lysine, a mild hemolytic anemia results. It seems almost incredible that a single amino acid residue in the wrong place, which does not cause any major changes in the protein structure, could have such a devastating effect on normal biological function and result in serious disease. The study of diseases caused by the alteration of molecular structure is a relatively new endeavor. It has great promise in understanding and subsequently conquering many types of diseases.

The results cited above show that the biochemical machinery that synthesizes protein does not always function at its optimum. To learn more about this process, as well as about the transmission of genetic information, we have to study the properties of the nucleic acids.

Nucleic Acids—General

It has been known for almost a century that hereditary information is transmitted from one generation to another by the chromosomes of the cells. The region of the chromosome that determines a particular characteristic or trait is called a gene. It has also been established that protein synthesis is controlled by the genes. A gene therefore possesses a dual function central to life: it must be able to replicate itself so that genetic information is transmitted to successive generations and must be able to direct the synthesis of proteins so that the organism can function. The fundamental chemical constituent of the genes is a long-chain nucleic acid, deoxyribonucleic acid, popularly called DNA. Although DNA is the carrier of the genetic information and contains the message directing the protein synthesis, the actual assembling of the appropriate amino acids and their polymerization is carried out by another class of polymeric nucleic acids, the ribonucleic acids, or RNA.

The underlying molecular basis for reproduction and protein synthesis lies in the chemical composition and structure of the nucleic acids. These vital biological functions will be seen to be natural consequences of the chain-like character of these polymers and the chemical nature of their repeating units. A schematic diagram of the nucleic acid chain is shown in Fig. 27. The backbone of the nucleic acid chain consists of repeating sugar units connected by phosphate bridges. Attached to each of the sugars is an organic base, either a purine or a pyrimidine. The major chemical distinction between the DNA and RNA chains is in the kind of sugar unit involved. Although in both cases the sugar is a five-membered ring, it is D-ribose in RNA and deoxy-D-ribose in

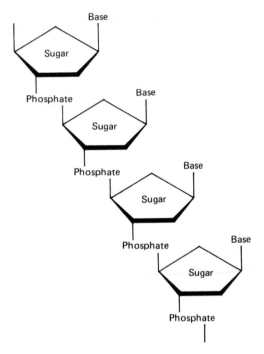

Fig. 27 Schematic representation of nucleic acid chain.

DNA, as is shown in Fig. 28. One sugar contains two hydroxyl groups while the other contains only one.

D-Ribose   Deoxy-D-Ribose

Fig. 28 Structural formulas for D-ribose and deoxy-D-ribose.

The purine bases attached to the sugars are either adenine (A) or guanine (G). Their structures are represented in Fig. 29.

The pyrimidine bases are cytosine (C), uracil (U), and thymine (T). Their structures are shown in Fig. 30. Cystosine and thymine are found in DNA, while in RNA uracil replaces thymine. The last two

130

Adenine (A)　　　　　　　　　　　Guanine (G)

Fig. 29　Structural formulas of purine bases.

bases differ by the presence or absence of a methyl —CH$_3$ group. In each of the nucleic acids, therefore, four chemically different bases are found. Figs. 29 and 30 also show that the pyrimidine bases are much smaller than the purines.

Cytosine (C)　　　　　　Uracil (U)　　　　　　Thymine (T)

Fig. 30　Structural formulas of pyrimidine bases.

Detailed skeletal structures of both types of nucleic acid chains are given in Fig. 31. The chain of nucleotides, as they are called, is perpetuated by a phosphate ester linkage. One can imagine the polymerization to take place by the reaction between phosphoric acid and the hydroxyl groups of the sugar. In addition to the usual conformational versatility that can be expected in all chain molecules, there is strong chemical specificity inherent in the nucleic acids through the composition and sequential distribution of the bases.

Deoxyribonucleic Acid—DNA

The molecular weight of the genetic nucleic acid, DNA, is extremely high. Depending on the species, the molecular weight can range from several million to several hundred millions. These molecular weights correspond to several thousand to hundreds of thousands of nucleotide repeating units. The extended length of a DNA molecule from the chro-

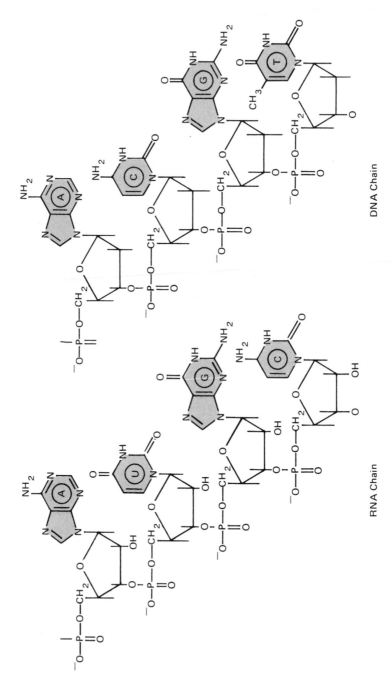

Fig. 31 Skeletal structure of representative portions of RNA and DNA chains.

RNA Chain

DNA Chain

mosomes depends on the complexity of the organism from which it comes. For example, it has been estimated that if the DNA in a human cell were stretched out, it would make a thread about three feet long. It is, however, so compacted in the cell that dimensions are actually in the submicroscopic range.

To be able to store and to transmit genetic information, the chain must be capable of having a code built into its structure. We must, therefore, seek to elucidate both the code and the mechanism by which it is transmitted from one generation to another. Since the chain backbone is made up of identical repeating units, it is difficult to see how any information can be stored or encoded in this part of the structure. However, there are four different bases attached to the sugar. Their concentration and sequential distribution are in fact the only variables in the chain structure. The possibility thus exists that if the bases were organized in definite sequences, these arrangements could constitute the genetic code. The very high molecular weight of DNA also suggests the capacity to store a great library of information.

The first substantial clue that the bases are actually involved in the coding process came from a very comprehensive chemical analysis of DNA. This analysis established some very general relationships for the base composition. It was found by Chargaff, from a study of DNA's from many sources, that the number of adenine (A) bases always equals the number of thymine (T) bases. Furthermore the number of guanine (G) bases is equal to the number of cytosine (C) bases. Consequently, the ratio of the total purine to the pyrimidine bases must be unity. On the other hand, the ratio $(A + T):(G + C)$ is highly variable. Thus there are two distinct, identifiable pairs of purine and pyrimidine bases. The amount of each of the bases making up a particular pair is the same. However, the relative amount of each of the pairs is variable. The universality of these relations for all DNA's makes it a quite significant observation and of underlying importance to DNA function.

The next problem that arises is how these compositional rules endow DNA with its special properties. The answer resides in the particular ordered structure of the DNA molecule. As we have seen from our discussion of synthetic polymers, polypeptides, and proteins, the details of ordered conformations can be obtained by the methods of X-ray diffraction. From the analysis of fiber patterns of highly oriented DNA, which were produced by M. H. F. Wilkins, Crick and Watson deduced their classical double-stranded helical model of DNA.[1] A space filling

[1] In their original paper, in 1953, Crick and Watson suggested that this structure had certain novel features of considerable biological interest. They in fact opened a major pathway of molecular biology, a scientific discipline that seems to explain biological function in terms of molecular structure.

molecular model of the Crick-Watson DNA structure is given in Fig. 32a and in a schematic, and perhaps somewhat more illuminating form, in Fig. 32b. The Crick-Watson model of DNA consists of two complementary chains that run in opposite directions and are coiled around each other to form a double helix. It gives the appearance of a spiral staircase, or twisted ladder. The sides or banisters are made up of the phosphate-sugar chain backbone. The steps or rungs consist of two bases, one from each chain, which lie in a plane perpendicular to the chain axis. The distance between the chains is such that one of the bases must be a purine (A or G) and the other a pyrimidine (C or T). A pair of purine bases is too large to fit into the allocated space while a pair of the smaller pyrimidine bases will leave a gap between the chains. We can see immediately that these requirements for the ordered structure exactly fulfill the demands of the base composition, namely that the number of A bases must equal the number of T while the number of C bases must equal the number of G. Thus each base pair, which forms the rungs in the ladder, is made up of either an A—T or C—G combination. This condition simultaneously satisfies both the steric and chemical requirements but still allows a great deal of freedom in the base sequences since there are no restraints on the ratio $(A + T):(G + C)$. The simliarity in the dimensions of these two allowable base pairs is shown in the diagram of Fig. 33. Besides satisfying the dimensional requirements, additional stability is given to the ordered DNA structure by these particular pairs by a set of very favorable hydrogen bond interactions which are indicated by the dashed lines.

When this structure is examined in more detail we find that the two polynucleotide chains are twisted around each other in such a way as to form a right-handed helix which completes one turn every 34 Å. The base pairs, looking like a set of stacked pennies, are spaced 3.4 Å apart so that there are ten base pairs per turn of the helix. The diameter of the molecule is such as to only be able to accommodate the two particular sets of base pairing that we have described. The double stranded molecule has a molecular weight of 196 per Angstrom unit of length or approximately two million per micron.[2] In higher organisms DNA molecules as long as 1,000 to 2,000 microns have been observed and correspond to fantastically high molecular weights. An electron micrograph of an elongated DNA molecule is shown in Fig. 34.

The chemical and structural requirements for specific base pairing are the key to the genetic function of DNA. Suppose, for example, that the four different bases were distributed along a single DNA chain in the manner indicated in the left-hand side of Fig. 35. However, since only particular bases can be linked with one another the sequence of

[2] One micron corresponds to 10,000 Angstroms or 10^{-4} cm.

Fig. 32 (a) Molecular model of ordered DNA structure (b) Schematic representation of ordered DNA structure. Reproduced with permission from E. J. DuPraw, *Cell and Molecular Biology* (1968) Academic Press, N. Y.

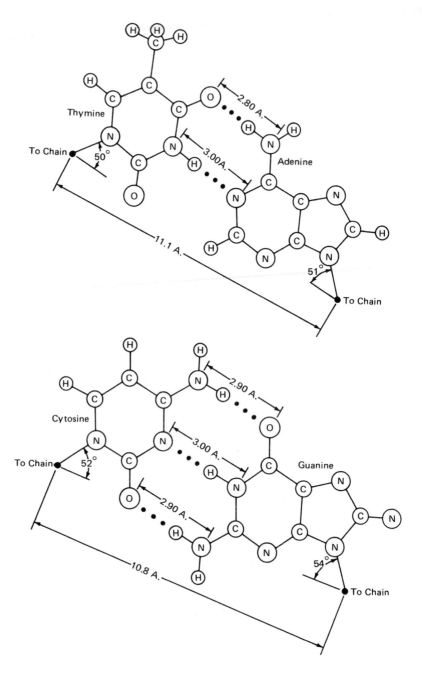

Fig. 33 Dimensions and hydrogen bonding capabilities of base pairs. Reproduced with permission from C. B. Anfinsen, *The Molecular Basis of Evolution,* John Wiley and Sons.

Fig. 34 Electron micrograph of elongated DNA. Reproduced with permission from E. J. DuPraw, *Cell and Molecular Biology* (1968) Academic Press, N. Y.

bases on the other chain in the molecule is immediately fixed by the sequence of the first chain. Thus in Fig. 35, A must be linked to T,

Fig. 35 Schematic representation of the complementary nature of the two chains in the DNA molecule.

T to A, G to C, T to A and so on down the line. One chain thus serves as a template for the base sequence of the other. Since the two chains are complementary, rather than identical, they can have widely different base compositions.

We can now visualize biological replication at a molecular level in a very definite way. When a cell is replicated it is necessary to provide the new daughter cell with an exact copy of the DNA molecule in the original mother cell. The beauty, and in some respects the simplicity, of the Crick-Watson structure is that each of the polynucleotide chains contains full information to exactly specify the base sequences of the other. Consequently if the two chains should separate each can form a template for the synthesis of a new complementary chain. The replication process is illustrated schematically in Fig. 36. The double-stranded parent molecule separates into two single chains. Each of these chains serves as a template for the synthesis of a complementary chain. Thus, two double-stranded daughter DNA molecules are formed which are identical in all respects to the parent. This process can of course be repeated for successive generations.

With replication the genetic message has been transmitted from one generation to the next in the four-letter GACT language. Although many of the details of the mechanism of strand separation and of template polymerization still remain to be elucidated the salient features of replica-

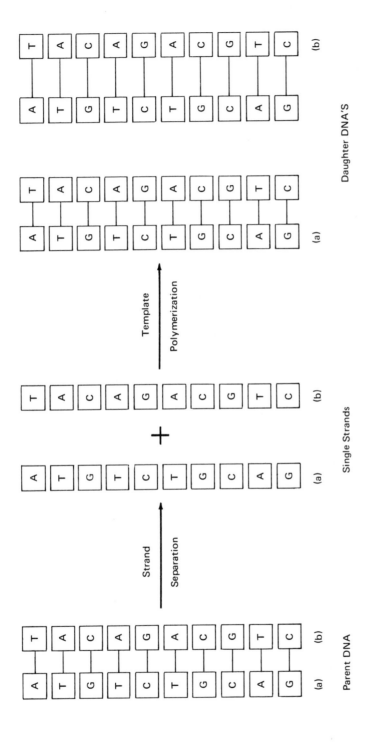

Fig. 36 Schematic representation of replication process.

tion have been established experimentally. Ingenious isotopic labeling experiments have enabled the progress of the individual chains to be followed through successive generations. The replication process has indeed been found to follow the pathway that has been outlined above. An enzyme that catalyzes the polymerization of DNA, from a pool of the four monomer units, has been isolated by Kornberg. However, for the polymerization to proceed a primer DNA must be present. This DNA presumably acts as a template for the polymerization since a complementary base composition to the primer is found in the new polymer formed.

We have seen how the structure of DNA enables it to perform one of its major functions, that of replication. For an organism to exist it must also have its complement of the many thousands of necessary protein molecules. Although these proteins are all different, they are all synthesized according to instructions contained in the DNA code. Put another way, the DNA molecule contains the information for assembly, in the proper order, of all the amino acids required to make one protein and has this information for all the proteins necessary to sustain living organisms. It has been estimated that the information stored in DNA for a human, if written out in the English language, would require the equivalent of several sets of a 24-volume encyclopedia. Although the information for protein synthesis is contained within the DNA molecule the extraction of this information and the actual protein synthesis is carried out in the cell through the intervention of the ribonucleic acids, RNA.

Ribonucleic Acids—RNA

Before examining the role of RNA in protein synthesis we should recall how it differs chemically from DNA. The sugars in the chain-backbone are different (see Figs. 28 and 31) and in RNA the pyrimidine base thymine (T) is replaced by uracil (U). Three very different kinds of RNA molecules have been identified as being involved in protein synthesis. Each of these is itself synthesized under the control of DNA. These RNA's are called messenger RNA, transfer or t-RNA and ribosomal RNA.

Messenger RNA accounts for about 5 to 10 percent of the total RNA in a cell. It is a short-lived molecule whose function is to receive the code from the DNA of the genes and to carry it to the ribosomes which are in the cell cytoplasm. The ribosomes are the site of protein assembly or the factory. Thus messenger RNA is both a transcriber and transmitter of the information for protein synthesis. Transfer RNA is a relatively low molecular weight molecule, composed of from 70 to

80 repeating units. Its function is to attach itself to a particular free amino acid, transport the amino acid to the synthesis site and to make sure that it is inserted in the proper place in the growing polypeptide chain. Ribosomal RNA, whose molecular weight is on the order of a million, together with certain proteins form the ribosomes. The ribosomes are small structures scattered throughout the cell cytoplasm. The message and the monomers are brought together here. RNA molecules do not possess the complementary base pairing characteristics of DNA. They exist as single stranded molecules and except for t-RNA do not exhibit any simple well-defined ordered structure. Yet, despite these apparent shortcomings, these molecules perform their assigned tasks admirably well.

Messenger RNA performs its task because it possesses a sequence of bases which are complementary to a portion of one of the DNA strands. This transcription process, from DNA to RNA, involves the polymerization of the RNA monomers in the correct order utilizing a DNA template. Within the cell nucleus, a portion of the DNA molecule unwinds and serves as a template for the polymerization of messenger RNA. The polymerization requires the active help of a specific enzyme. For example, a DNA base sequence of A—T—C—G—T will be transcribed into U—A—G—C—A in a messenger RNA. The sequence of bases in DNA determines the sequence of bases in RNA. In the transcribed molecule the bases control the order in which the amino acids are incorporated into a growing polypeptide chain. This is the genetic code in action. The question then arises as to what specifically is the message.

The operation of the genetic code must obviously involve a relationship between the four bases on the nucleic acids and the 20 possible amino acids used in making a protein. Since there are only four different bases there cannot be a one to one correspondence between a base and an amino acid. Similarly, since there are at most 16 different nucleotide pairs in sequence, a direct correspondence between pairs of nucleotides and all amino acids must also be ruled out. The coding for a single amino acid residue must, therefore, contain at least a sequence of three nucleotides. It turns out that the code is a triplet or three-letter word such as AUC. Since there are 64 combinations of triplets, (4^3), there must be a redundancy, or degeneracy as it is called, in the coding. Before examining the specifics of the genetic code we summarize in Fig. 37 the role of DNA and messenger RNA in the storage and transcription process. The necessary information is stored in the two stranded base-paired DNA molecule in the sequence of bases. The message is transcribed onto messenger RNA using the first strand of the DNA. Each successive grouping of three bases on the messenger RNA, called

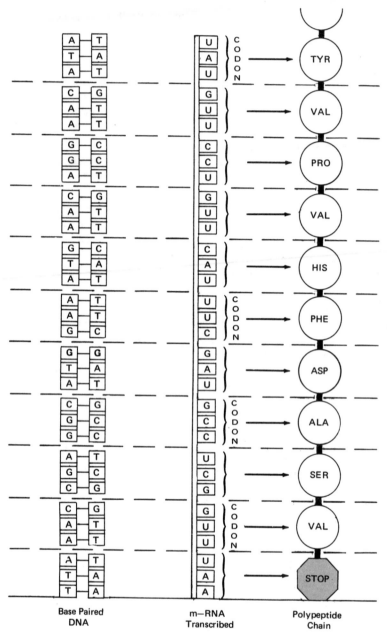

Fig. 37 Schematic diagram of storage and transcription role of the nucleic acids in directing syntheses of the last ten residues of ribonuclease.

a codon, codes for the particular amino acid indicated. In this example, we have actually shown the last 10 amino acid residues of ribonuclease (see Fig. 12).

In a set of ingenious experiments from the laboratories of Nirenberg, Oochoa, and Khorana, which will have far-reaching implications for mankind, the direct correspondence between the base triplets and the amino acid residues has been established. The results of these experiments are summarized in Table III.

Table III
Genetic Code: Relation Between Base Triplets and Amino Acids

Amino Acid	Triplet or Codon Assignment					
Alanine	GCC	GCU	GCA	GCG		
Arginine	CGC	CGU	CGA	CGG	AGA	AGG
Asparagine	AAU	AAC				
Aspartic Acid	GAU	GAC				
Cysteine	UGU	UGC				
Glutamic Acid	GAA	GAG				
Glutamine	CAA	CAG				
Glycine	GGU	GGC	GGA	GGG		
Histidine	CAU	CAC				
Isoleucine	AUU	AUC	AUA			
Leucine	UUA	UUG	CUU	CUC	CUA	CUG
Lysine	AAA	AAG				
Methionine	AUG					
Phenylalanine	UUU	UUC				
Proline	CCC	CCU	CCA	CCG		
Serine	UCU	UCC	UCA	UCG	AGU	AGC
Threonine	ACU	ACC	ACA	ACG		
Tryptophane	UGG					
Tyrosine	UAU	UAC				
Valine	GUU	GUC	GUA	GUG		
Terminate (Stop)	UAA	UAG	UGA			

All 64 possible triplet combinations are accounted for. Some of the amino acids are coded for by as many as six different triplet combinations. On the other hand methionine and tryptophane are each coded by single combinations AUG and UGG respectively. Also included in the table are the code words for the termination of the growing chain, i.e., the instruction to stop the polymerization when the protein assembly has been completed. We note that there is no code word for two of the amino acid residues found in proteins—hydroxyproline and cystine. These residues must obviously be synthesized from their close relatives, proline and cysteine, after the polypeptide chain is formed. Thus the covalent disulfide cross-links typical of many proteins, and involving residues well removed from one another along the chain, are formed after the protein is put together.

Before discussing protein synthesis further, we shall examine the consequences of mistakes in the primary code of the DNA or in its transcription. An error in the positon of one of the bases, i.e., if it were incorrect or misplaced, results in a different triplet, so that a different amino acid residue in a particular position in the protein results. These changes are usually brought about accidentally in nature by the action of high energy radiation or by chemical agents. Because of these errors, changes or mutations are found in the proteins of the progeny. In some cases these mutations have only a minor effect, but in others they can be disastrous. As an example we have already noted that in sickle cell anemia the glutamic acid residue in the sixth position in the β-chain of hemoglobin is replaced by valine. As Table III shows, the codons for glutamic acid are GAA and GAG. If the middle base is replaced by U the resulting triplets GUA or GUG codes for valine. We see then that this superficially small change is the genetic basis for the particular molecular disease. Its cure must involve correcting the DNA message. Sometimes the message is so garbled in a particular gene that an important enzyme cannot be synthesized. This can cause a major disruption of metabolic processes and very often leads to early death or mental retardation among infants.

Returning to the problem of protein synthesis, we have already pointed out that the monomers are brought to the polymerization site by t-RNA. The transfer RNA molecule must not only select the proper amino acid from the monomeric pool but must also be able to read the code on the messenger RNA so that the amino acid is placed in the proper position in the growing chain. All the t-RNA molecules contain about 70 to 80 repeating units. In addition to the usual RNA bases, others are found in small but significant proportions. There is at least one particular t-RNA for each amino acid; different amino acids do not become attached to the same t-RNA molecule. In some cases, however, each of several different t-RNA molecules can select and attach itself to the same amino acid.

Several of the t-RNA's have now been purified and the detailed sequence of their bases has been determined. However, their three-dimensional structure has not as yet been worked out in detail. Hence we can only surmise the broad outlines of how they function by correlating their chemical structure with the acts they perform. All transfer RNA's have a common sequence of terminal bases, namely C—C—A. The sugar (ribose) unit associated with the terminal adenine base has both of its hydroxyl (OH) groups free. As is shown in Fig. 38 the amino acid is attached to one of the OH groups. The specificity of

Fig. 38 Terminal bases of t-RNA.

amino acid selection involves the intervention of an enzyme particular to the amino acid. The molecule must also have a sequence of three bases, which are not hydrogen-bonded to any others, and which are complementary to the codon of the messenger RNA. This structural feature will allow the amino acid to be inserted in the proper place. For example, if methionine is to be transferred, then the codon on the messenger RNA calling for this amino acid is AUG. Therefore, the complementary sequence, or anticodon, UAC capable of hydrogen bonding with AUG must be present in the t-RNA. In this way the t-RNA will find its proper match along the messenger RNA. The minimal structural requirements of this t-RNA are represented in highly schematic form in Fig. 39.

Fig. 39 Schematic structure of methionine t-RNA.

When the t-RNA has acquired its amino acid it diffuses to the ribosomes. Here, after appropriate interaction with the messenger RNA protein synthesis occurs. The ribosomes are roughly spherical particles about
200 Å in diameter. They are composed of both RNA and protein and
have a molecular weight of about 2.7×10^6. Ribosomal RNA does not

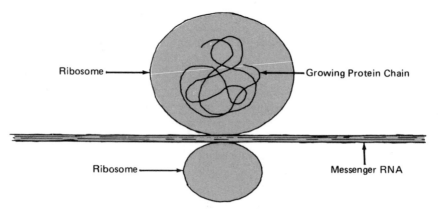

Fig. 40 Schematic representation of ribosome interacting with messenger
RNA.

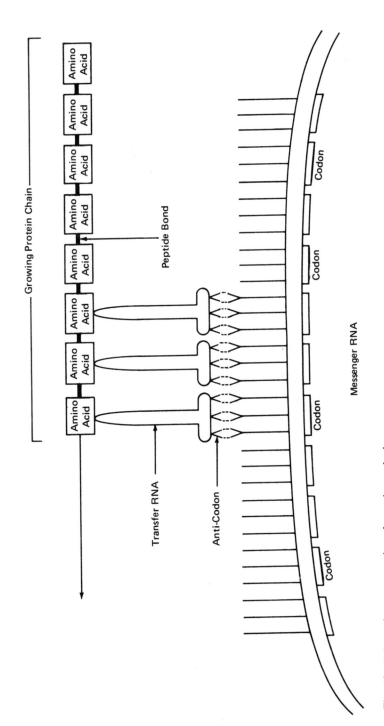

Fig. 41 Schematic representation of protein synthesis.

contain genetic information and the details of its role in protein synthesis are still obscure. The ribosome is composed of two distinct subunits one being about twice the size of the other, as it shown in Fig. 40. The protein synthesis does not take place in homogeneous solution but is carried out on the surface of the ribosomes. The chief function of the ribosome is to orient the t-RNA and its associated amino acid with the messenger RNA so that the code can be read accurately. Ribosomes must thus contain specific surfaces that bind t-RNA, messenger RNA, and the growing polypeptide chain in proper orientation.

A schematic representation of the polymerization is given in Fig. 41. The t-RNA and the growing polymer chain appear to be attached to the larger of the ribosomal subunits; whereas messenger RNA is bound to the smaller one. The growing chain is retained by the ribosome until synthesis is complete. There is a great deal still to be understood about details of protein synthesis and the specific role played by the ribosomes. It is known, however, that the polypeptide chain is initiated and grows from the N-terminal amino acid end in a stepwise manner. As each repeating unit is added, the ribosomes move in a fixed direction relative to the messenger RNA. This motion allows the next codon to interact with a t-RNA. The amino acid at the other end of the t-RNA is thus in

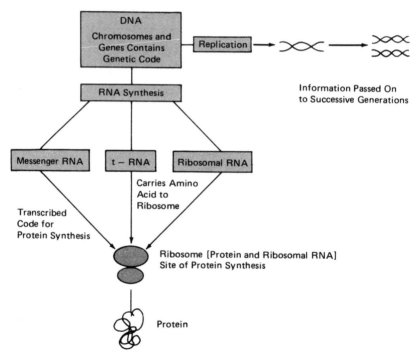

Fig. 42 Summary diagram of role of nucleic acids in biological function.

position to be properly attached to the peptide chain. After the "stop" codon is reached, the chain is released by the ribosome. The sequence of amino acid residues determines the proper three-dimensional structure of the protein. The time required for the synthesis of a polypeptide chain containing 300 to 400 amino acid residues is about 10 to 20 seconds.

Before concluding we summarize, by means of Fig. 42, the diverse roles of the nucleic acids in biological function.

EPILOGUE: MACROMOLECULES AND MAN

We have seen from our discussion of macromolecular structure and properties the features that are common to these substances as well as the diversity of functions they perform and the many uses to which they are put. Their essential nature to life as we know it today, both as a biological necessity as well as to our comfort and well being, is readily apparent. With the fundamental understanding that has evolved regarding this class of molecules we can anticipate, with great confidence, the development of whole new classes of polymeric substances as well as the laboratory synthesis of many biopolymers that are crucial to life processes. The history of science has taught us to expect that much more will yet be coming. The present, as well as the anticipated advances in polymer science will have a profound impact on our society as it influences our standard of living, our ecology and environment and, through molecular biology and molecular biophysics, our control of biological processes. This ever-increasing scientific bounty will bring with it, however, many problems and responsibilities. The solution of these problems and the proper and conscientious assumption of these responsibilities rests with all of society. However, intelligent solutions cannot be devised without a real scientific understanding of the molecular nature of the systems with which we are concerned. If we have accomplished this latter task in this book, then it is appropriate to briefly outline and consider the major problems that are already with us or that can be easily anticipated in the very near future.

Synthetic materials, particularly plastics, are often thought of as being cheap, inferior products of our highly technological Western society. It is argued that we can easily do without such materials. In an absolute sense this may be true for a limited number of products. In a relative sense, however, this concept is clearly not warranted. The properties of most synthetic materials equal or surpass those provided by nature. If we were to depend solely on natural sources for these materials they would soon become depleted because of the demand, even if the earth's population remained constant. This depletion would have devastating effects on our environment since our forests would be stripped and our land eroded. Those of us who live in affluent societies, where we take our shelter, clothing and transportation for granted could undoubtedly very easily dispense with a great deal of our new materials and luxury. This is not so, however, for those who live in poorer societies. More than half of the earth's population is made up of the undeveloped countries of the world. For those people, what we consider nonessentials can easily serve as a base for a substantial improvement in the quality of their life. Here the use of substitute synthetic materials offers by far the greatest opportunity to enhance their standard of living and bring it nearer to a level that we have come to expect and enjoy. By taking advantage of the technological advances that have been made in polymer science and of the large quantities of polymeric materials that can be produced very economically they can be provided with many of the comforts of life.

For example, when one starts virtually from scratch without concern for traditional methods of production and vested interests, then housing and clothing and tires for transportation can be produced economically and can be widely distributed. Poly-*cis*-isoprene, natural rubber, can now be manufactured in chemical plants. The plantation-based production and economy can be eliminated and the commercial products derived from this polymer could be made more widely available. Synthetic fiber plants, strategically situated around the world, could supply clothing for all. Major advances can also be made in housing. The use of polymer-fiber glass reinforced materials or polyurethane foam for the housing shells makes possible sturdy but inexpensive and rapid construction. This concept in housing is not a theoretical dream but is actually being realized in a very modest way today. This effort only has to be expanded to bring decent shelter to everyone. Furnishings made from the ABS type copolymers can make living more pleasant and comfortable. These few examples indicate the comforts that can be provided that would otherwise be virtually unattainable. Polymer technology should thus be directed to the benefits of humanity so that its great promise will be fulfilled. It is our responsibility as members of society to make sure that such efforts are vigorously pursued.

There are, then, very compelling reasons for the production of synthetic materials and the continued research and quest for new and better products. If we accept this principle, then we must also realize that it presents a very serious problem and challenge. The problem in its simplest form is that of waste disposal. Although many naturally occurring polymers such as cellulose and natural rubber will degrade with time, i.e. are biodegradable, synthetic materials in the form of fibers, containers and packaging material must be disposed of. We certainly do not want a major portion of the earth surface to become covered with polymeric debris. Unlimited production must eventually lead to this situation since polyolefins, such as polyethylene and polypropylene, and vinyl polymers, such as polystyrene and polyvinyl chloride, are not biodegradable. They cannot be recycled as paper, glass and certain metal containers can, nor does the development of recycling processes appear technologically feasible in the foreseeable future. Is the solution to this vexing problem the limitation of production either indirectly by taxation, as has already been attempted in some cities, or directly by legal restrictions? If so, this obvious and very simple approach would mean denying the benefits of the technological advances in substitute and very utilitarian material to those segments of our civilization whose standard of living is low and can be easily raised. This dilemma can be resolved if we take advantage of the fact that most of the polymers will burn. What is required to solve the problem, then, is the development of incinerators that are self-contained and thus will not cause pollution of the atmosphere. Although such incinerators have not been developed they are certainly technically feasible and represent a way to protect our environment while still making available the benefits of new materials to the people who need them the most. Much effort should thus be given to the development of such incinerators. Serious problems do not usually have simple solutions. We must be concerned with the impact on all of civilization rather than the small segment of society in which we find ourselves when we decide on a course of action.

Another set of very serious responsibilities faces us in the light of the revolution in molecular biology that has occurred in our lifetime. Many far-reaching implications are inherent in the discoveries that have already been made. For the first time in his history man understands his chemical origin and function. He can now investigate and understand life at its most fundamental molecular level and has the capability, in principle, to control the design and fate of future generations. DNA has already been synthesized in the laboratory as have certain genes and a completely functioning enzyme. In the not too distant future, man will have the capability to create man.

The promises to mankind that can be derived from this knowledge are very great. Cure and control of genetic diseases, detection and rectification of genetic errors so that our progeny will not be abnormal or deformed all lie within our grasp. The molecular structure of viruses, their control and relation to cancer and other diseases will be unraveled before long. These benefits, coming as they will from our understanding of the structure and function of biological macromolecules, justifies the effort, the time, and frustration of the basic scientific research that underlies all these efforts.

Accompanying all the good that we can foresee there are also many danger signals. It is no longer in the realm of science fiction to anticipate the creation of life in a test tube and the synthesis of living organisms.

Is this to go on uncontrolled, left to individuals' desires and passions? Major experimental efforts designed to understand brain function, memory, and the learning process are now under way in many laboratories. Proteins and nucleic acids are involved in these functions, and we can expect a fundamental understanding of these processes to emerge shortly. Clearly these are no longer scientific or technological problems but problems that must be the deep concern of society as a whole. How we address ourselves to these matters and control our actions may very well decide whether man will survive or become another extinct species in evolutionary history. It is not too early for us to become deeply involved and concerned.

The acceptance of the macromolecular hypothesis and the consequences of scientific study have given us materials unthought of 30 years ago and a detailed understanding of the fundamentals of life processes. We stand on the threshold of a new era in our civilization. Whether the Age of Macromolecules will be man's golden age or a disastrous period clearly rests with man himself.

SUGGESTED BIBLIOGRAPHY FOR
ADVANCED READING

For a more detailed discussion of polymer properties, at about the same level as this monograph, consult

R. B. Seymour, *Introduction to Polymer Chemistry,* McGraw-Hill Book Company

The classic treatise in the field, which is a superb exposition, is

P. J. Flory, *Principles of Polymer Chemistry,* Cornell University Press

This book is very strongly recommended for those who wish to pursue the study of polymers in more detail. A knowledge of undergraduate physical chemistry is, however, necessary for a full understanding.

There are several monographs on specialized subjects. They can be read with the background that has been developed here. These are:

L. R. G. Treloar, *Physics of Rubber Elasticity,* Oxford University Press, Oxford
L. Mandelkern, *Crystallization of Polymers,* McGraw-Hill Book Company, New York
J. D. Ferry, *Viscoelastic Properties of Polymers,* John Wiley and Son, New York
R. W. Lenz, *Organic Chemistry of Synthetic High Polymers,* Interscience Publishers, New York

In the area of biopolymer structure and the properties and function of proteins and nucleic acids further suggested reading can be found among:

R. G. Dickerson and I. Geis, *The Structure and Action of Proteins,* Harper and Row, New York
J. D. Watson, *Molecular Biology of the Gene,* Second Edition, W. A. Benjamin, Inc., New York
E. J. DuPraw, *Cell and Molecular Biology,* Academic Press, New York

There are also films available. These are:

Physical Chemistry of Polymers, prepared by Bell Telephone Laboratory. Available from Sterling Movies Inc. 43 West 61st Street, New York, New York 10023.
Chemistry of the Cell. I: The Structure of Proteins and Nucleic Acids II. Function of DNA and RNA in Protein Synthesis. Available from the McGraw-Hill Book Company. New York.

Index

Hooke's Law, 56, 59
Howells, E. R., 45
Hydrogen bond, 41, 69, 102, 116, 137, 145
Hydrophillic groups, 100, 119, 121
Hydrophobic groups, 100, 119, 121
Hydroxyapatite, 114
Hydroxyproline, 115, 118, 119, 143

Initiation, 21
Insulin, 109
Intermolecular interactions, 63, 103
Internal rotation, 25, 26, 27, 28, 29, 46, 54, 63, 108
Isometric force, 94
Isoprene, 68
Isotactic polymers, 40, 47, 49, 83, 103

Kendrew, J. C., 126, 127
Keratin, 109, 111, 112, 113
Khorana, H. G., 143
Knobler, C. M., 120
Kornberg, A., 140

Lamella crystals, 74, 75, 77
Lenz, R. W., 154
Lexan, 11
Ligaments, 114
Lucas, F., 115
Lucite, 9
Lysozyme, 120, 121, 122, 123, 124

Mandelkern, L., 72, 74, 154
Marsh, R. E., 106
Mechanical properties, 61
Mechanochemistry, 89
Melamine, 23, 24
Melting temperature, 81, 82, 83, 84, 85, 88
Membranes, 108
Methane, 13
Methionine, 143, 145, 146
Molecular biology, 133, 150, 152
Molecular biophysics, 150
Morphology, 71, 73, 74, 75, 77, 81, 88
Muscle, 94, 95, 108, 111, 114, 124
Mutation, 144
Mylar, 11
Myoglobin, 124, 126, 127
Myosin, 111

Natta, G., 47
Natural rubber, 8, 10, 40, 57, 61, 62, 63, 64, 68, 73, 83, 84, 151
Neoprene, 10

Network
    deformation, 57, 58
    formation, 22, 23, 24, 57, 58
    three dimensional, 6, 7, 56
Neutrons, 24
Nirenberg, M., 143
Nucleation, 69
Nucleic acids
    deoxyribonucleic acid (DNA), 129, 130, 131, 132, 133, 138, 144, 152
    general, 1, 12, 24, 41, 96, 129, 131, 146, 149
    messenger RNA, 140, 141, 144, 145, 146, 148
    ordered structure, 133, 134, 135
    ribonucleic acid (RNA), 124, 129, 130, 132, 140
    ribosomal RNA, 140, 141, 146
    transfer (t) RNA, 140, 144, 145, 146, 148
Nylon, 10, 24, 41, 42, 43, 83, 84, 85, 89

Ochoa, S., 143
Orlon, 9
Oth, J. F. M., 90

Paraffin, 13
Pauling, L., 106, 112
Permanent waving, 112
Perutz, M. F., 127
Phenol-formaldehyde polymer, 23
Plastic flow, 51
Plasticizers, 64, 65
Plastics, 1, 17, 24, 84, 150
Plexiglas, 9, 61
Polyacrylonitrile, 9, 83, 84, 85
Polyadenylic acid, 12
Polyalanine, 12, 101
Polyamides, 10, 20, 40, 41, 43, 51, 64, 69
Poly α-amino acid, 12, 40
Polybutadiene, 10, 64
Poly carbonate, 11
Poly chloroprene (neoprene), 10, 64
Polydimethyl siloxane (silicone rubber), 40, 63, 64, 83, 84
Polyester, 10, 20, 41, 51
Polyether, 10
Polyethylene
    branched, 4, 82, 83
    general properties, 2, 13, 17, 25, 26, 30, 32, 33, 34, 38, 39, 40, 45, 67, 73, 74, 76, 78, 79, 84, 151
    linear, 4, 82, 83